Investigation and Prevention of Officer-Involved Deaths

Investigation and Prevention of Officer-Involved Deaths

Cyril H. Wecht, JD, MD
Forensic Pathologist

Henry C. Lee, PhD
Professor, University of New Haven

D.P. Van Blaricom
MPA, Chief of Police (Ret)

Mel Tucker
MPA, Chief of Police (Ret)

Illustrations:
Scott G. Roder/Roder Evidence Consulting

CRC Press
Taylor & Francis Group
Boca Raton London New York

CRC Press is an imprint of the
Taylor & Francis Group, an **informa** business

CRC Press
Taylor & Francis Group
6000 Broken Sound Parkway NW, Suite 300
Boca Raton, FL 33487-2742

© 2011 by Taylor and Francis Group, LLC
CRC Press is an imprint of Taylor & Francis Group, an Informa business

No claim to original U.S. Government works

Printed in the United States of America on acid-free paper
10 9 8 7 6 5 4 3 2 1

International Standard Book Number: 978-1-4200-6374-5 (Hardback)

Library of Congress Cataloging-in-Publication Data

Investigation and prevention of officer-involved deaths / Cyril H. Wecht ... [et al.].
    p. cm.
Includes bibliographical references and index.
ISBN 978-1-4200-6374-5 (hardcover : alk. paper)
1. Police misconduct--United States. 2. Homicide--United States. I. Wecht, Cyril H., 1931- II. Title.

HV8141.I58 2011
363.2'2--dc22   2010038208

Visit the Taylor & Francis Web site at
http://www.taylorandfrancis.com

and the CRC Press Web site at
http://www.crcpress.com

Table of Contents

8   In-Custody Deaths   157

9   Emotionally Disturbed Persons   173

Preface

A great philosopher once stated that he could readily determine the level and extent of any civilization by studying the way in which police officers function and the manner in which penal institutions are conducted. It is with that maxim in mind that the authors undertook to compile the contents of this book.

Each year, approximately 150 law enforcement officers die while performing their duties. For the first part of this century, the majority died when engaged by suspects with firearms. In addition to officer deaths, there are approximately 375 people killed each year by the police. Between 2000 and 2008, there were 3,000 people killed by the police who were classified by the Federal Bureau of Investigation (FBI) as justifiable homicides. Over 98% of those were shot to death with a firearm.[1]

In recent years, more officers have died from accidents while on duty than by a felon with a firearm. Approximately 36% of all officer line-of-duty deaths in recent years have been vehicle-related. In addition, hundreds of innocent bystanders die each year when involved in a collision with a police car or fleeing vehicle while operating their vehicle, or when struck by a police car or fleeing vehicle while walking. Although there is no national database, it is estimated by the National Highway Traffic Safety Administration (NHTSA) that the police are involved in about 70,000 police chases each year resulting in the deaths of 400 innocent people.[2]

Most people would agree that police officers whose daily assignments require them to be directly engaged in various kinds of potentially dangerous situations have a grave responsibility to protect innocent bystanders as well as themselves. In our democratic society, it is also incumbent upon the law enforcement official to utilize only as much potentially lethal force as necessary regarding those individuals who are pursued, arrested, and incarcerated for obvious, alleged, or suspected crimes. A valid and logical argument can be made that there is no dichotomy between these two pragmatic objectives. The physical safety and well-being of the uninvolved third party and the police officer most often flow in a parallel fashion with the nonviolent apprehension of the "actor."

Of course, there are numerous scenarios in which the police officer has no alternative but to employ the full force of his or her weaponry in pursuing, arresting, and subduing a dangerous, violent individual. Regrettably, these kinds of situations result in the tragic deaths of many law enforcement officers every year in the United States. Unfortunately, many deaths of completely innocent third parties and potential victims also occur as a consequence of such violent confrontations between police and suspected criminals.

What about the larger number of people whose deaths annually are directly or indirectly related to the actions of law enforcement officials during pursuit, apprehension, arrest, and incarceration? Should our society ignore these deaths and simply attribute all of them to unchallengeable professional decisions and necessary acts of police? Or does it behoove an advanced civilization like ours to objectively review all such police-related deaths in a diligent, thorough, open, and unbiased fashion in order to determine what the circumstances were that ultimately resulted in that individual's death?

Analysis and examination of police-related deaths are not solely undertaken to retrospectively ascertain whether the involved officer acted in a deliberately improper or unintentionally negligent fashion. Such reviews are certainly intended to accomplish that objective, which is necessary for moral, ethical, and legal reasons. However, there is a more overarching purpose for these kinds of postmortem case studies—namely, the continuing, advanced education of all active-duty police officers and other law enforcement officials so that they can better serve themselves and the society in which they function.

The authors created several representative scenarios of officer-involved deaths from their extensive professional experiences. The examples that follow describe circumstances wherein law enforcement officers have had to respond to critical incidents that are outside of their usual experience and have unfortunately resulted in the death of the person they were trying to take into custody. It is the authors' intent to show that an examination of such incidents will serve to better inform law enforcement practitioners on how to thoroughly investigate officer-involved deaths and thereby learn how they might have been prevented.

The major and most frequently occurring kinds of police-related deaths are set forth on a chapter-by-chapter basis. In each category, a hypothetical, quite realistic scenario is presented. With such a background scenario in place, relevant discussions then follow dealing with the important and critical issues that need to be considered and evaluated from the perspective of law enforcement officials, criminalists, forensic pathologists, and other forensic scientific experts.

One of the most important areas in investigation of police-related deaths is the crime scene investigation and the collection and preservation of the relevant forensic evidence to prove or disprove certain issues and hypotheses. We have outlined the appropriate crime scene procedures for each situation and listed the potential categories of forensic evidence that should be searched for, collected, and sent to the forensic (crime) laboratory for analysis. Those laboratory analysis procedures are set forth in many laboratory manuals and textbooks, and therefore are not included in this book.

Postmortem protocols, consisting of autopsy reports, toxicological analyses, and other appropriate investigative findings depicting particular types of police-related deaths are included.

Studies of police-related deaths have shown that restrictive policies in high-risk police activities save lives. Accordingly, we included a chapter dealing with policy and training.

It has also been demonstrated that police use of less-lethal weapons can save lives. Hence, we included a separate chapter that discusses less-lethal weapons.

The authors pose the critical question—"What would you do if you were the police officer?"—in each of the scenarios that are presented to challenge the reader.

The authors believe that a serious study of each of the categories presented in this book will enable police officers and other groups of law enforcement officials to more fully comprehend and appreciate the societal significance of such cases. The more aware, sensitive, and well-educated that officers of the law are, the more dignified, humane, and safe the communities that they are sworn to serve will be.

No law enforcement officer wants to be involved in a death that could have been avoided by a better understanding of a particular dynamic with which he or she may be suddenly confronted. Accordingly, a thorough investigation will lead to a better understanding of what has occurred and encourage improved training for future similar events. The authors understand that not all officer-involved deaths can be prevented. However, we believe that some can. The information provided in this book should be helpful in achieving that highly desirable goal.

We greatly respect the difficult duties that our dedicated law enforcement officers perform and hope to make some of their dangerous encounters safer for them.

Endnotes

1. U.S. Department of Justice, Federal Bureau of Investigation, *Crime in the United States*.
2. Voices Insisting on Pursuit SAFETY (VIPS) (www.pursuitsafety.org).

About the Authors

Cyril H. Wecht, J.D., M.D., is one of this country's leading forensic pathologists. He received his medical degree from the University of Pittsburgh School of Medicine. He also holds a law degree from the University of Maryland School of Law. As a medical expert, he has performed more than 17,000 autopsies and has reviewed or supervised over 36,000 additional postmortem examinations. Dr. Wecht is a former president of the American College of Legal Medicine and the American Academy of Forensic Science. He is a fellow of the College of American Pathologists and the American Society of Clinical Pathologists. He has served as a medical-legal and forensic pathology consultant in civil and criminal trials since 1962.

Henry C. Lee, Ph.D., is one of the world's foremost forensic scientists. He has worked on most of the challenging cases that have occurred during the past 40 years, including the O.J. Simpson trial, the JonBenet Ramsey investigation, the suicide death of White House Counsel Vincent Foster, and the reinvestigation of the John F. Kennedy assassination. Dr. Lee was the Commissioner of Public Safety for the State of Connecticut and was their Chief Criminalist from 1979 to 2000. He is currently Chair Professor of Forensic Science University of New Haven (Connecticut) and Director of The Henry C. Lee Research and Training Center (also at the University of New Haven).

D. P. Van Blaricom began his law enforcement career in 1956 as a patrol officer with the Bellevue Police Department (Washington). He later served as a detective, sergeant, captain, and deputy chief before being selected as the Chief of Police in 1975, where he served until his retirement in 1985. Since his retirement from active law enforcement, Van Blaricom has served as a litigation consultant in over 1,500 law enforcement cases throughout the United States.

Melvin L. Tucker began his law enforcement career as an FBI Agent in 1969 and served as a police chief for four cities in three states before retiring from active law enforcement service as the Chief of Police for the City of Tallahassee, Florida, in 1994. Since his retirement, Tucker has served as a litigation consultant in over 450 law enforcement cases throughout the United States.

Reducing and Preventing Deaths by Training and Policy Guidance

1

Law enforcement has become a big business. There are now 17,876 state and local law enforcement agencies in the United States employing 731,903 officers.[1] The total direct expenditure for federal, state, and local law enforcement has now reached $185 billion annually.[2] Because law enforcement is now a big business and is routinely involved in arresting and detaining people; engaging in vehicle pursuits; using force to overcome resistance to arrest; and searching homes, people, and vehicles, there are officers, suspects, and innocent citizens who suffer injury and death when suspects resist violently or poor tactics and judgment are used by officers. During the past century, over 14,000 federal, state, and local law enforcement officers were killed in the line of duty. Of those killed, 49% were shot to death, making it the single leading cause of officer deaths.[3]

In addition to officer deaths, approximately 375 people are killed each year by the police. These deaths end up being classified by the Federal Bureau of Investigation (FBI) as justifiable homicides. Over 98% of the justifiable homicides involve people who were shot to death with a police firearm.[4]

Vehicle-related activities (motorcycle or automobile accidents or struck by vehicle) were the second leading cause of death of law enforcement officers over the past century, accounting for approximately 30% of all officer line of duty deaths.

In addition, hundreds of innocent bystanders are killed each year when struck by or involved in a collision with a police car or fleeing vehicle. Although there is no national database, it is estimated by the National Highway Traffic Safety Administration (NHTSA) that the police are involved in about 70,000 police chases each year, resulting in the deaths of 400 innocent people.[5]

High-Risk Business

Police officers wear uniforms and drive cars with emergency lights and agency markings. As a consequence, their activities are noticed by the public. When they make mistakes, they, and their employing agency, are often subjected to lawsuits seeking monetary relief for the harm incurred.

One of the most frequent allegations made against law enforcement officers is that they used unnecessary or excessive force in carrying out their duties. In 2002, there were 22,238 citizen complaints against municipal law enforcement officers alleging unnecessary or excessive force; 2,815 against sheriff's deputies;

763 against county police officers, and 740 against state police agencies.[6] Those complaints often evolved into lawsuits. Although there is no national data collection effort focused on lawsuits filed against the police, it is estimated that the number of lawsuits filed against law enforcement officers has tripled in recent years with an estimated 30,000 new lawsuits being filed in state and federal courts each year against law enforcement agencies and their officers.

Civil Liability

In the past, law enforcement officers, supervisors, and municipalities were protected from civil liability by the doctrine of sovereign immunity, but today, officers, supervisors, and municipalities may be held liable if inadequate or improper training causes injury or violates a person's civil rights. This liability may be based on the concept of negligence under a state law and filed in state court or a civil rights violation filed in federal court.

Negligence

Negligence is the breach of a legal duty owed to a person that causes injury. The duty owed to a person may be breached by any action that falls below the reasonable standard of care customarily exercised by members of the law enforcement profession. The duty owed to a person may also be breached by a failure to act where there is a legal duty to act, such as the failure to provide adequate training.

Civil Rights

Federal civil rights law imposes liability on police officers, police supervisors, and municipalities for violating any person's civil rights. A failure of an agency to train a police officer, or a failure of an agency to provide adequate training, may serve as the basis for holding a municipality liable under 42 U.S.C. 1983.[7] However, the municipality will only be liable when the failure to train amounts to a deliberate indifference to the rights of a person with whom the police come into contact.[8]

Law Enforcement Training

Unfortunately, the law enforcement profession has been slow to develop adequate training programs. The idea for the first formal training program for law enforcement officers originated with August Vollmer, the father of modern law enforcement in the United States. The school started in 1908 with officers attending during their off-duty time, and it consisted of training on police methods, photography, first aid, and criminal law.[9] The next year, the New York City Police Department established a formal "academy" to provide recruits with training in firearms, rules and regulations of the department, police procedures, and criminal law.[10]

In 1930, the Wickersham Commission, established to study law enforcement in the United States, recommended the standardization and professionalization of law enforcement agencies through centralized training.

In response to the Wickersham Commission recommendation, the FBI, in 1935, moved to exert its influence more directly on local law enforcement when it started the FBI National Academy for the training of municipal, county, and state police training officers. The regular instructors at the FBI National Academy were experienced and highly trained. The FBI also made a practice of bringing in nationally prominent educators, criminologists, and attorneys to teach specialized courses. Graduates generally agreed that the program was the best in the nation. Unfortunately, the 12-week program only admitted 80 officers per class, and only two classes were given each year.[11] Even today, the FBI Academy admits only 250 officers four times a year to its academy located at Quantico, Virginia.[12]

The International Association of Chiefs of Police (IACP), founded in Chicago in 1893, had as its original goal the apprehension and return of criminals who had fled the agency jurisdictions in which they were wanted.[13] Over the years, the IACP expanded into advancing the science of police work by promoting improved training and practices. In particular, the IACP Training Key Program, consisting of a series of semimonthly training bulletins and instructor guides, was inaugurated in March 1964. Each Training Key was designed to be a self-contained lesson plan discussing a topic of importance to police officers. Today, more than 600 Training Keys have been made available to law enforcement agencies on topics ranging from how to make a felony vehicle stop to the use of deadly force. Law enforcement agencies in all 50 states currently participate in the program.[14]

In 1967, in response to widespread concern over the quality of law enforcement services in the United States, the President's Commission on Law Enforcement and Administration of Justice was formed resulting in the Omnibus Crime Control and Safe Streets Act being passed by Congress the following year. That legislation created the Law Enforcement Assistance Administration (LEAA) to administer federal funding to state and local law enforcement agencies. Before it was abolished in 1982, LEAA funded law enforcement educational programs, research, and local crime initiatives. One of the initiatives funded by the LEAA was the formation of the National Advisory Commission on Criminal Justice Standards and Goals.

In 1973, the National Advisory Commission on Criminal Justice Standards and Goals published its *Report on Police* recommending, among other things, that having an effective police service in the United States required the establishment of mandatory minimum basic training for the police and the creation of agencies to develop and administer training standards and programs.[15]

Contemporary Law Enforcement Training

Today, all states mandate a minimum number of hours of training for all new police officers. The typical state academy requires approximately 600 hours of training. Some states also mandate a minimum number of hours of annual in-service training. The number of hours of training actually provided new officers is directly related to the size of the employing agency, with large agencies (those serving populations of 1,000,000 or more) generally providing 1000 hours or more of preservice training, and the smaller agencies (those serving populations of 10,000 or less) generally providing only the minimum state-mandated number of hours of training.[16]

Basic Recruit and In-Service Training Programs

Currently, there are two types of ongoing police training programs: basic recruit training and in-service training programs. In-service training (roll-call or specialized classroom instruction) is designed to constantly update the skills and knowledge of veteran officers.

Recruit training is the initial training an officer receives to provide the officer with the minimal necessary skills and knowledge required to perform the job of a police officer. It should be emphasized that basic training provides only the minimum skills and knowledge to perform the job but does not test the recruits' ability to perform the skills in a field setting while under stress or in rapidly changing circumstances.

The Field Training Officer Program

Since the first formal field training officer (FTO) program was established in San Jose, California, in 1972, many police departments throughout the United States have initiated FTO programs. The FTO program is a formalized method of evaluating new officers. This is accomplished by assigning the new recruit graduate from the basic training academy to work under the supervision of an FTO. The FTO trains the new officers by allowing them to apply the principles on the street that they learned in the classroom during basic recruit training at the academy. The FTO program has become tremendously important in ensuring the quality of the police recruit training program. FTO programs are designed to ensure that recruits have the basic competencies to perform as police officers in a field setting while under stress.[17]

Job Task Analysis and Training

Today's basic recruit, in-service, and FTO programs are all designed to provide training for officers based upon the core skills necessary to perform the task of a law enforcement officer as determined by a job task analysis. Although it is necessary to train all officers on the tasks they are likely to perform while in the field, if we are to reduce police-involved deaths and injuries, additional training programs need to be provided to all officers that focus on the areas in which most loss of life and injuries occur.

The Need for Change and Leadership

The size, complexity, and risk associated with providing law enforcement services make it extremely important that police chiefs, sheriffs, and directors of all law enforcement agencies accept the responsibility of determining what skills and knowledge their officers need in order to perform their jobs. And then put into place training programs to provide them with the necessary skills and knowledge. Although much improvement in the quality and quantity of police training has been realized during the past two decades, much still needs to be done.

Training to Prevent Mistakes and Liability

Implementing highly focused (directed toward those activities that account for the greatest percentage of officer, suspect, or innocent person deaths) and effective training programs, spelling out very clearly what it is students are to learn, accompanied with an evaluation tool to determine how well they learned it, is essential to reducing mistakes and officer-involved deaths.

A well-organized training program, with the appropriate documentation of that training, is not only one of the only ways to reduce the number of mistakes officers make in carrying out their duties, but is also one of the most effective ways of reducing liability.

Focused Training on Authority to Use Force

In 1985, the U.S. Supreme Court rendered a decision regarding the use of deadly force by a law enforcement officer in arrest situations. In *Tennessee v. Garner*, 471 U.S. 1, the Court stated that a law enforcement officer's use of deadly force was a "forcible seizure" and therefore subject to the "reasonableness" requirement of the Fourth Amendment.

The effect of *Tennessee v. Garner* was to limit the use of deadly force by a law enforcement officer to (1) the protection of himself or herself or another from a threat of serious bodily harm or death, or (2) to prevent the escape of a person who may justifiably be characterized as "dangerous" to others if allowed to remain at large.

Since *Garner,* law enforcement officers have been told that they have the authority to use deadly force to protect themselves or another from a threat of serious bodily harm or death, but they have not been told that their perception of that threat must be reasonably made. As a consequence, lawsuits alleging that officers were unreasonable in their threat assessment and took preemptive action when it was not justified have been increasing.

In 1989, the U.S. Supreme Court extended the "reasonableness" requirement of the Fourth Amendment to an officer's use of both deadly and nondeadly force in both arrest and detention situations when they rendered their decision in *Graham v. Connor*, 490 U.S. 386.

Because the Court reaffirmed its position that all police use of force incidents should be analyzed under the Fourth Amendment "reasonableness" standard, the effect of *Graham v. Connor* was to establish the authority of law enforcement officers to use force, in arrest or detention situations, to protect themselves or others from an immediate threat of serious injury or death as long as the threat assessment was reasonable, and the level of force used in response was reasonable.

The Court, in *Graham v. Connor*, demonstrated a high level of consideration for law enforcement officers and the difficulty involved in decision making in the field when it declared the following:

> The reasonableness of a particular use of force must be judged from the perspective of a reasonable officer on the scene, and the calculus of reasonableness must allow for the fact that police officers are often forced to make split-second judgments, in circumstances that are tense, uncertain and rapidly evolving, about the amount of force that is necessary in a particular situation.[18]

Focused Training on Preemptive Action

Many deaths of citizens today involve officers shooting a suspect they claim was reaching for a weapon but subsequent investigation determined the suspect was unarmed. The concept of taking action because an officer believes a threat has been recognized, but before the threat has been carried out, has become known in the law enforcement profession as "preemptive" action.

Officers frequently find themselves responding to incidents that require them to consider whether or not to take preemptive action to prevent a serious injury or to save a life. However, they are not being trained on situational awareness and how to exercise good judgment when making their threat assessment.

The reasonableness of the officer's perceptions (threat assessment) must be considered in light of the nature of the offense, the events that preceded the offense, and the apparent dangerousness of the suspect (the totality of the circumstances known to the officer) before the reasonableness of the officer's response (threat mitigation) can be determined.

For example, officers are taught in use of force training programs that noncompliance with verbal commands should be considered as a danger signal when making a threat assessment. There is nothing wrong with that training as a general rule.

However, officers also need to be told that a threat assessment, to be reasonably made, requires consideration of the parameters of the incident, including the preceding events, the exposure of the officer, the proximity between the parties involved, the likelihood of a weapon being present, the capability of that weapon, and any other relevant objective factors. They need to be told that any officer taking preemptive action solely on the basis of noncompliance with verbal commands is likely to make mistakes and find himself or herself in the courtroom defending his or her actions.[19]

As an example, assume the following set of facts: (1) an officer stops a suspect in a car that was reported to have been used in a bank robbery 2 hours earlier; (2) bank employees reported that they noticed during the robbery the suspect was armed with a .45-caliber semiautomatic handgun; and (3) the suspect, after being stopped, refused to follow the officer's commands to show his or her hands. Under this set of facts, it would be entirely reasonable for the officer to interpret the failure of the driver to follow his or her commands as threatening behavior. If the refusal to follow the officer's commands is coupled with any furtive movements by the subject, the officer would be authorized under the law to use deadly force because his or her perception of a danger of serious bodily harm or death was reasonable under the set of facts and circumstances confronting him or her.

However, assume this set of facts: (1) an officer stops a vehicle because the officer observed the driver weaving across traffic lanes and driving without headlights on at night; (2) the officer checked with dispatch before the stop and they advised that they had no reports of the vehicle being reported as stolen; (3) dispatch also advised the officer the vehicle was registered to a local citizen; (4) dispatch advised the driver's name was checked and came back with no record; and (5) after the vehicle stopped, the driver refused to follow verbal commands from the officer to show his or her hands.

Absent any other information to suggest that the suspect was armed or dangerous in any way, an officer taking preemptive action and using deadly force under this set of facts and circumstances would likely find that his or her action would not pass the test of reasonableness, because the driver's failure to follow verbal commands was consistent with someone being so intoxicated the driver did not understand what the officer was ordering him or her to do.[20]

Policies Can Reduce Deaths

In New York City, an 11-year-old African American male was shot and killed by the police in 1972. The community uproar that followed the shooting prompted the police commissioner to implement a new firearms policy that restricted the use of deadly force to those situations that involved a defense of life. Many of the largest 50 cities in the United States followed the example set by the New York City Police Department (NYPD). A later study of shootings between the years 1971 and 1984 within those cities with populations over 250,000 found a large decrease in shootings, with the police killing 172 citizens in 1984 compared to 353 citizens in 1971.[21]

Supreme Court Decisions on the Use of Force

As mentioned previously, in 1985, the U.S. Supreme Court handed down its decision in *Tennessee v. Garner* (471 U.S. 1) which restricted the circumstances under which a law enforcement officer may use deadly force to make an arrest. It was the first time the Court had addressed police use of force. The *Garner* case involved the death of a 15-year-old African American male who had broken into an unoccupied house and stolen $10 cash from a coin purse. Memphis police officers called to the scene saw Garner run toward the rear of the house surrounded by a six-foot-high chain-link fence. The officers shined a flashlight on Garner, saw his hands, and were "reasonably sure" that he did not possess a firearm. As Garner started over the fence, the officer, believing that he would escape, fired at Garner, hitting him in the back of the head and killing him. The shooting conformed to the Tennessee statute on use of force and the Memphis Police Departments' policy in effect at the time of the shooting.

Interestingly, the U.S. Supreme Court in *Garner* used the fact that many large law enforcement agencies had already restricted their officers' use of deadly force to "defense of life" situations to allay fears that their decision would be problematic for the law enforcement profession. The Court decision that "deadly force may not be used unless it is necessary to prevent the escape and the officer has probable cause to believe that the suspect poses a significant threat of death or serious physical injury to the officer or others" had little real effect on the largest law enforcement agencies that had been operating under that standard for many years.

Some who study police policy and practices have suggested that we still lose officers to the same mistakes that we have always lost officers to because we are not remediating persistent problems associated with officer decision-making skills.[22] Historical data confirm that they are right. A review of the law enforcement officers feloniously killed in the line of duty from 1996 through 2005 shows that 25% of all officers killed were slain during arrest situations, 18% in ambush situations, 12% while investigating suspicious activities, and 3% during tactical situations.[23] However, we continue to train officers on skills and glamorous activities like defense against edged weapons and special weapons and tactics instead of officer safety during arrest, during vehicle operations, responding to calls for service, and situational awareness to avoid ambush. As stated by police trainer Thomas Aveni, "Contrary to prevailing police training emphasis, poor judgment gets more officers killed, seriously injured and sued far more frequently than poor marksmanship does."[24]

Policy Guidance and Domestic Violence

For many years, the common practice of the police when responding to domestic violence calls was to avoid arrest because (1) many officers believed that domestic fights were private matters, (2) many officers believed that the female victim would end up being uncooperative if an arrest of her assailant were made, and (3) many officers believed that arresting the breadwinner would hurt the family. However, in 1984 the results of the Minneapolis Domestic Violence Experiment were released.[25] The results demonstrated that restricting police officers' discretion on whether to arrest, and requiring arrest where probable cause existed, would produce a deterrent to future domestic violence. The study, which covered a period of 18 months, found that when the police did not make an arrest, the nonarrested offender was twice as likely to reoffend as the offender who was arrested by the police. This was one of the first studies to show that arrest worked to reduce violence and that restricting officer discretion could also play a role in reducing violence.

The Exercise of Discretion

Law enforcement officers are faced with numerous situations every day that require them to make quick decisions, usually without time for input from a supervisor. For example, when making a traffic stop, the officer will need to make a decision as to whether to use a legal sanction, such as arrest, or whether to issue a ticket. If the officer decides to make an arrest, and the traffic offender decides to resist arrest, the officer will then have to decide on whether to use physical force, and how much force, to overcome the offender's resistance. When the officer makes his or her decision, the officer is choosing among a number of possible courses of action. The officer's choice is an exercise in police discretion.[26]

Prevention of Illegitimate Exercise of Discretion

An officer's exercise of discretion is not a problem as long as the officer does not consider the offender's race, color, gender, creed, national origin, ethnicity, ancestry, religious beliefs, age, marital status, sexual orientation, or physical disability when exercising his or her judgment. Simply stated, the exercise of discretion is a necessary part of law enforcement and not problematic unless the exercise of judgment is made on an illegitimate basis.

National Advisory Commission on Criminal Justice Standards and Goals

The National Advisory Commission on Criminal Justice Standards and Goals, appointed by the Law Enforcement Assistance Administration, in its 1973 Report on Police recommended that every police agency should adopt "comprehensive policy statements that publicly establish the limits of discretion, that provide guidelines for its exercise within those limits, and that eliminate discriminatory enforcement of the law" (Standard 1.3, page 21). The Commission Report also recommended that every police chief should formalize procedures for developing and implementing the policy statements.[27] One of the purposes of the Commission's recommendation was to prevent law enforcement officers from considering illegitimate factors when exercising their discretion; however, the adoption of policy statements and procedures for developing and implementing those policies is not a simple task.

Variables

Research has broken down the decision elements in the exercise of discretion into three categories:

1. *Offender variables*—Adult complaints are taken more seriously by the police than those made by juveniles, persons who respond to the police with deference are treated more leniently, arrest and use of force is more likely to be used against non-Caucasians, and persons in upper-income brackets receive better and more services from the police.

2. *Situation variables*—Serious matters (crimes) get more attention from the police than do minor matters (noncrime); activities initiated by the police are followed up with greater intensity than activities initiated by citizen complaint; and vice crimes are not enforced until citizens complain, the vice activity becomes commercialized, or the activity becomes highly visible (the three Cs of complaints, commercialization, and conspicuousness).

3. *System variables*—The police are less likely to initiate arrest when the courts and correctional systems are clogged; police tend to initiate arrest when city revenues are down and revenue is generated by arrest; and police are less likely to arrest when their communities have strong social services such as mental health facilities and detox centers.[28]

Although it has become accepted in the law enforcement profession that the exercise of police discretion is the essence of police work and indispensable, it has also become accepted that there must be properly confined, properly structured, and properly checked policy and procedure guidance to limit unrestricted and unnecessary discretion.

Policies and Procedures

Policy is different from rules and procedures. Policy should be stated in broad terms to guide employees. It sets limits of discretion. A policy statement deals with the principles and values that guide the performance of activities, directed toward the achievement of agency objectives. A procedure is a way of proceeding—a routine—to achieve an objective.[29]

The National Law Enforcement Policy Center

In 1987, the IACP, in cooperation with the U.S. Department of Justice, established the National Law Enforcement Policy Center. The purpose of the National Law Enforcement Policy Center was to advance professional police services through promoting best practices in both administration and operational matters. As of October 2008, the National Law Enforcement Policy

Center had published over 100 *Model Policies* on topics ranging from a law enforcement officer's off-duty conduct and HIV/AIDS in the workplace to dealing with the mentally ill and dealing with domestic violence. Each of the *Model Policies* is accompanied with a *Concepts and Issues Paper* that provides essential background material and supporting documentation to provide for a greater understanding of the developmental philosophy and implementation requirements for the model policy.[30]

Police Policies and Official Immunity

Generally, under the theory of official immunity, when an officer exercises discretion—that is, exercises the officer's professional judgment to choose from alternative courses of action—there is no liability unless the evidence shows that the officer engaged in a willful or malicious wrong.[31] This creates an issue of balance to be considered by the police administrator. On the one hand, the more discretion left to an officer, the stronger is the defense of official immunity when a claim is filed against an agency. On the other hand, leaving too much discretion to the individual officer may result in a higher risk of harm to citizens or a loss of the official immunity defense, and may possibly create a legal duty or standard of conduct if language such as "shall" or "will" is in the policy statement. However, whether or not a department policy can create a legal duty is a question the U.S. Supreme Court has not directly addressed.[32]

Identification of Areas in Which Policy and Procedure Guidance Is Needed

Officer conduct that will likely result in the injury or death of the officer, suspect, or innocent citizens, if improperly performed, should be the type of conduct on which policy and procedure guidance should focus. One high-frequency and high-risk activity is police pursuit and emergency response driving. Another high-frequency and high-risk activity is use of all levels of force. These two activities account for the majority of injuries and deaths of suspects and innocent citizens. These activities require policy and procedure guidance and extensive officer training if officer-involved deaths and injuries are to be reduced or prevented.

Summary

We must continue to train officers on the core activities performed by the typical patrol officer as shown by a job task analysis (JTA), and we must continue to verify the recruit officer's ability to perform the core activities in a field setting through use of FTO programs. However, if we are to reduce the number of officer-involved deaths, training of law enforcement officers in the future must be directed to the areas in which there is the greatest likelihood of officer, suspect, and innocent person deaths occurring. That means that we are going to need to focus training of officers on the use of force as a preemptive action, violence-reduction strategies, emergency vehicle operations, safety in traffic control, safe driving of motorcycles, and better tactics when performing the core activities of arrest and handling disturbance calls.

In addition, restricting police officer discretion in some high-risk areas, by providing the officer with policy and procedure guidance, can reduce mistakes, save lives, and reduce the frequency of civil litigation.

Police leaders must identify those areas of officer core performance requirements that generate the highest incidence of injury, death, and litigation and determine whether policy and procedure guidance (along with training on those procedures) can reduce or prevent mistakes.

It is no longer an issue as to whether officer discretion will be reduced, but in what areas will the reduction in discretion benefit both the officer and those the officer comes into contact with while performing his or her duties.

Endnotes

1. U.S. Department of Justice, Bureau of Justice Statistics, Census of State and Local Law Enforcement Agencies, 2004, *NCJ Bulletin* 212749, June 2007.
2. U.S. Census Bureau, Annual General Finance and Employment Survey, 2003.
3. National Law Enforcement Officers Memorial Fund, Washington, DC.
4. U.S. Department of Justice, Federal Bureau of Investigation, Crime in the United States.
5. Voices Insisting on Pursuit SAFETY (VIPS) (www.pursuitsafety.org).
6. U.S. Department of Justice, Bureau of Justice Statistics, *Citizen Complaints about Police Use of Force*, June 2006.
7. *Canton v. Harris*, 489 U.S. 378 (1989).
8. *Monell v. New York City Department of Social Services*, 436 U.S. 658 (1978)
9. Alfred E. Parker, *Crime Fighter: August Vollmer*, New York, McMillan, 1961.
10. *Police Department, City of New York: Self Portrait*, July 1959.
11. Allen Z. Gammage, *Police Training in the United States*, Springfield, IL, Charles C Thomas, 1963.
12. Federal Bureau of Investigation (www.fbi.gov/hq/td/academy/na/na.htm).
13. "International Association of Chiefs of Police," Wikipedia (http://en.wikipedia.org/wiki/International_Association_of_Chiefs_of_Police).
14. IACP Training Key, Volume Five, Introduction, page 2.
15. National Advisory Commission on Criminal Justice Standards and Goals, *Report on Police*, 1973, page 384.

16. U.S. Department of Justice, Bureau of Justice Statistics, *Local Police Department Training*, 2002.

17. L. K. Gaines, M. D. Southerland, and J. E. Angell, *Police Administration*, pages 279–283, 1991.

18. *Graham v. Connor*, 490 U.S. 386 (1989).

19. Melvin Tucker and Chris Wisecarver, Legal Authority for Pre-Emptive Action, *National Tactical Officers Association Magazine*, Spring 2008.

20. Ibid.

21. Ibid.

22. Thomas J. Aveni, *Obsolescence: The Police Firearms Training Dilemma*, Spofford, NH, The Police Policy Studies Council, 2008.

23. Ibid.

24. Ibid.

25. L. Sherman and R. Berk, The Specific Deterrent Effects of Arrest for Domestic Assault, *American Sociological Review,* 49, 261–272, 1984.

26. Kenneth Culp Davis, *Police Discretion*, Eagen, MN, West, 1975.

27. Ibid. at page 101.

28. Austin Peay State University (http://www.apsu.edu/oconnort/4000/4000lect07.htm).

29. National Advisory Commission on Criminal Justice Standards and Goals, *Report on Police*, 1973.

30. See, as an example, IACP National Law Enforcement Policy Center, Use of Force, Concepts and Issues Paper 2-05.

31. Ann Gergen, *Police Policies and Official Immunity*, League of Minnesota Cities, 2007.

32. John C. Hall, Liability Implications of Departmental Policy Violations, *FBI Law Enforcement Bulletin*, April 1997.

Less-Lethal Weapons

2

Tools that can be used by law enforcement officers to stop, control, and restrain individuals while causing the least possible harm to the individual and, at the same time, not increasing the danger to the officer are now generally available to contemporary law enforcement officers; however, there has been an ongoing debate in the law enforcement profession for the past few years over what to call these tools. Some have suggested these tools be called *weapons,* while others believe they should be referred to as *devices* under a theory that *devices* sounds better to the public than *weapons.* Some have argued that the law enforcement profession should call these *less-lethal weapons,* others *less-than-lethal weapons,* and still others *nonlethal weapons* under a theory that what these weapons are called can be important in the event of civil litigation over their use. The authors know of no such litigation occurring. Whether these tools of law enforcement are called devices or weapons or are called less-lethal, nonlethal, or less-than-lethal weapons does not really matter. All of these descriptors mean the same thing—any device that, when used in the appropriate manner, is designed and intended to bring a suspect under control without inflicting death or serious injury. For purposes of the discussion in this chapter, we will call these tools *less-lethal weapons,* because that has emerged as the most commonly used descriptor. Examples of less-lethal weapons include pepper spray, pepper balls, water guns, nets, batons, beanbags, rubber projectiles, and conducted energy devices (CEDs).[1]

Less-lethal weapons are now commonly utilized by local, state, and federal law enforcement officers. Data gathered in 2003 suggested that 98% of all local law enforcement agencies authorized their officers to use pepper spray when appropriate; 95% authorized the side-handle, collapsible, or traditional baton when appropriate; 28% authorized soft projectile weapons (beanbag) when appropriate; 23% authorized conducted energy devices (stun guns, TASERs) when appropriate; 8% authorized rubber bullets when appropriate; and 1% authorized high-intensity light weapons when appropriate.[2] More recent data suggest that there has been a large increase in the number of agencies authorizing their officers to use conducted energy devices. According to the Police Executive Research Forum (PERF), 8,000 police and sheriffs' departments across the country have now adopted the use of conducted energy devices.[3]

Reducing Injury and Death

The International Association of Chiefs of Police (IACP) has concluded that there is enough anecdotal evidence to show that when officers appropriately use less-lethal weapons to defuse potentially deadly situations, the number of injuries and deaths—for officers and others—decreases significantly.[4]

Types of Incidents

The type of incident in which less-lethal weapons are most utilized is suicidal persons who force the police to shoot them. This phenomenon, which has become known as suicide-by-cop, or SbC, accounts for 50% of the total incidents in which less-lethal weapons are used, followed by barricaded suspects at 24% of the total.[5]

Studies indicate that 30% of all SbC incidents can be successfully resolved through the use of less-lethal weapons, while only 17% of SbC incidents are likely to be successfully resolved when employing the strategy of negotiating with the suicidal person.[6]

The data for these studies were obtained in the 1990s when an officer's choice of a less-lethal weapon was very limited. At the time the data were collected, beanbag weapons and pepper spray were the weapons most widely available. In the late 1990s, conducted energy devices (electronic control weapons) were increasingly finding their way into law enforcement agencies and became the less-lethal weapon of choice by most officers. Because of the effectiveness of CEDs ability to immediately incapacitate a person, it is believed likely that future studies will show even greater success in saving lives in SbC incidents by the use of less-lethal weapons.

Less-Lethal Not Required by Law

Although less-lethal weapons can assist officers in gaining control in potentially violent confrontations and provide them with an alternative to the use of deadly force, their use is not required when the use of deadly force is justified.[7] The courts have consistently ruled that deadly force is justified any time a suspect threatens an officer with a weapon intending to inflict serious physical harm, and officers need not use less-lethal weapons under those circumstances.

However, a strong motivation for the development of less-lethal weapons has been the concern over deadly force lawsuits.[8] In addition, another motivating factor has been the many cases in which a suicidal person, armed with a knife and slowly advancing toward officers, was shot and killed by the police. In these cases, jurors have not been convinced the actions of the police met the "reasonableness" requirement and returned a verdict in favor of the plaintiff.

The Ideal Less-Lethal Weapon

Even though the use of less-lethal weapons is not legally required when deadly force is justified, officers, seeking an alternative to the use of deadly force, have been asking for them for many years. Research is now ongoing to determine what less-lethal weapons are effective and what negative side effects exist when they are used. The IACP with support from the Office of Community Oriented Policing Services, the Bureau of Justice Assistance, the National Institute of Justice, and others developed a clearinghouse for information on less-lethal weapons (www.less-lethal.org).

Specifically, the law enforcement profession has expressed a need for stand-off less-lethal weapons that:

- Are practical under the circumstances.
- Will cause immediate and temporary incapacitation.
- Will have minimal medical implications.
- Will give a high probability of immediate control.
- Will have observable effects.
- Will only affect the intended.
- Will have a low probability of causing death if misused.

Practicality is generally a matter of the location of the suspect (oleoresin capsicum [OC] is not as effective outdoors in windy conditions, capture nets are not practical when the suspect is moving, and none of the less-lethal weapons are practical when the suspect will be killed or injured if he or she falls). The immediate incapacitation of the suspect is a requirement because most deadly attacks against police officers occur spontaneously, cannot be anticipated, and occur at a close range. Temporary incapacitation is necessary to give the deploying officer enough time to safely approach the suspect and to restrain him or her before the effects from the less-lethal weapon wear off. Observable effects confirm for the officer that the device, or agent, deployed has had the desired effect and it is safe to approach the suspect. Finally, the probability of the device causing serious injury or death must be very low.[9]

The National Law Enforcement Technology Center

Law enforcement officers have had a very limited and largely inadequate set of less-lethal weapons to choose from. To remedy that problem, the National

Law Enforcement Technology Center (NLETC) has been working to develop stand-off less-lethal weapons. It is believed that as effective and reliable less-lethal weapons become more readily available to officers, and they are better trained on their appropriate use, the need for the use of deadly force will be reduced. In addition, there has been pressure by some states to limit law enforcement use of deadly force even when legally justified (particularly in dealing with emotionally disturbed persons who have not committed a crime). Some states have even developed law enforcement polices that prohibit the use of deadly force against any person whose conduct is injurious only to himself or herself.[10]

However, the only less-lethal weapons in the development stage or currently available include:

1. Thermal guns that raise body temperature so fast that they incapacitate.
2. Shotguns that fire nets.
3. Strobe lights that temporarily blind.
4. Darts tipped with drugs.
5. Motion-restraint foam guns.
6. Beanbags filled with various substances and fired by a standard shotgun.
7. Pepper sprays and pepper ball guns (paintball guns).
8. Special guns that fire wooden, plastic, or rubber pellets.
9. Guns that fire darts into suspect and deliver an electric shock.

Of the available less-lethal weapons, only beanbag guns (and a few other impact projectile weapons like the six shot, rotary cylinder, rifled barrel Sage Gun), pepper sprays and pepper balls, capture nets, and electric shock weapons are considered by the law enforcement profession to be of practical use in a law enforcement setting. Unfortunately, all of them have problems that have resulted in controversy and litigation.

Capture Devices

These weapons either shoot out a net similar to those used to capture wild animals or dispense sticky foam intended to immobilize a suspect. However, even though the net launchers work well outdoors on a stationary suspect, they are not practical for use in confined spaces, on moving suspects, or indoors. Glue guns, which shoot foam that expands up to 30 times its pressurized volume up to a distance of 35 feet, can be lethal if the foam covers a suspect's nose and mouth.

Impact Weapons (Beanbags)

Extended-range impact projectiles (beanbags) are designed to incapacitate dangerous persons from a distance to minimize the danger to the officers involved. In order to do this, they must deliver sufficient energy to overcome the suspect's resistance while not delivering so much energy that death or serious injury is likely to result.

Impact projectiles (beanbag rounds) are projectiles enclosed in heavy cloth and filled with pellets that can travel at approximately 280 feet per second. The effective range is about 50 feet. At distances of less than 10 feet, extreme caution must be exercised due to a high possibility of injury or death. At distances of 10 to 20 feet, caution must be exercised to avoid hitting the head, neck, heart, spleen, liver, and kidney areas.

However, "beanbag" weapons, even when deployed in accordance with the manufacturer's recommendations, (1) often will not incapacitate large individuals, (2) if fired at a short range will penetrate and injure, and (3) are notoriously inaccurate at longer ranges because of the low velocity of the impact projectile.[11]

In recent years, efforts have been made to improve the accuracy of the extended-range impact weapons by employing flexible tails to beanbags to provide stability in flight, by providing rigid, or semirigid, fins to steer and stabilize projectiles, or by adding ridges and grooves to the launching device to impart spin stabilization in flight. Although rifled barrels improve accuracy and reduce the fatality risk involved with beanbag use, the head and neck area, along with the heart area, are still vulnerable to fatal blunt trauma and need to be avoided.[12]

Pepper Sprays (Oleoresin Capsicum)

OC is a naturally occurring substance derived from the cayenne pepper plant. On contact with OC, the mucous membranes in the nose, throat, and eyes become inflamed and swollen, resulting in constricted airways and shortened breath, involuntary eye closure, coughing, and nasal drainage. OC, in the form of stream or spray, is employed widely by law enforcement agencies as a less-lethal weapon to accomplish crowd or individual person control. Although OC's inflammatory properties make it an effective less-lethal weapon and 90% effective in incapacitating suspects in human confrontational encounters,[13] it has not been consistently effective against violent, intoxicated, drugged, and mentally ill individuals. The primary problems associated with the use of OC have been (1) the possible contamination of innocent bystanders; (2) its occasional ineffectiveness when utilized against the emotionally disturbed person (EDP) or those under the influence of drugs; (3) the time necessary for incapacitation to occur; (4) the diversion of the OC carrier by wind or rain; and (5) the protection provided by eyeglasses, hands, and collars on clothing which can keep the agent from reaching the face and eyes.

The time necessary for OC to take effect depends on the age of the suspect (older people absorb OC more slowly than do younger individuals) and the carrier used for the OC (water, as a carrier, evaporates slowly, resulting in a longer time to take effect, while the use of isopropyl alcohol as a carrier reduces the time to take effect because the alcohol evaporates quickly). Typically, a compressed gas, such as isobutene, propane, or carbon dioxide, is used to propel the OC. Some propellants are flammable in a gaseous state, and caution must be used when deploying in certain circumstances. OC deployment can be effective up to 15 feet depending upon the width of the spray employed and the propellant used. Full recovery from the effects of OC can be expected within 45 minutes after exposure to fresh air and flushing of the eyes with water.

Although, to date, OC has not caused any deaths,[14] there have been deaths associated with the use of OC or attributed to pepper-spray pellet guns when used improperly during crowd control efforts.[15]

Conducted Energy Devices (CEDs)

CEDs have become the preferred terminology to describe weapons that employ electrical current (Figure 2.1). CEDs are also called electromuscular disruption devices (EMDDs) and electronic control devices (ECDs).[16]

The first CED was developed by Jack Cover, a National Aeronautics and Space Administration (NASA) scientist. Cover experimented with electricity as a nondeadly weapon in the 1960s. He discovered that when short-duration (milliseconds), high-energy, direct-current electric pulses were applied to human beings, immediate incapacitation almost always occurred with direct, negative side effects. This discovery led to a delivery system he called the TASER, an acronym for the Thomas A. Swift Electric Rifle. Swift was a

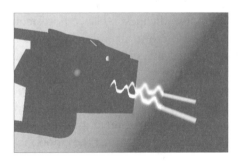

Figure 2.1 A close-up view of a conducted energy device (CED). (**See color insert following page 78.**)

fictional character in a book published in 1911 written by Victor Appleton. Cover spent several years perfecting this futuristic device, and it was introduced to the public through the 1976 Clint Eastwood film *The Enforcer*. The original electronic control weapons were 50,000-volt, seven-watt stun systems that were classified as a firearm, because they used gunpowder to fire probes into targeted subjects, and fell under the provisions of the 1968 Gun Control Act.

Later research and development efforts resulted in the introduction of newer electronic control weapons in 1999. The newer devices were 50,000-volt, 26-watt systems. Unlike their predecessors, the new versions used nitrogen cartridges, rather than gunpowder, to fire the probes. The device was classified as an electromuscular disruptor because it overrides the central nervous system. These weapons continued to be called TASERs.

TASERs can penetrate up to 5 centimeters of clothing, causing an uncontrollable contraction of the muscle tissue and immobilization regardless of pain tolerance or mental focus. TASERs are regarded, in law enforcement circles, as the only less-lethal weapon currently available that is effective and reliable in meeting the primary law enforcement objective of immediate and temporary incapacitation with minimum medical implication.

TASER® has become the largest electronic control weapon manufactured in the United States. The pistol-shaped M-26 and X-26 models are now used by more than 13,000 law enforcement, correctional, and military agencies around the world. Both the M-26 and X-26 produce electrical shocks that are delivered either through firing a dart that stays connected to the TASER with very thin wires (muscular disruption mode) or through touching the suspect with the device and delivering a shock (drive stun mode).

However, because these weapons are so reliable and effective, allegations have been made that many law enforcement agencies use these devices as a routine way of controlling uncooperative people, and TASERs may contribute to a person's death, especially if that person has an underlying medical condition or is under the influence of alcohol or drugs.

The medical effects of CEDs continue to be a matter of controversy because so many people have died shortly after being exposed to CED activations. These deaths have become known in the law enforcement profession as *proximity* deaths. The controversy is spurred on by both empirical data (there have been approximately 100 reported deaths associated with the TASER since 1999) and daily news stories of apparent TASER misuse.[17]

Recent data, however, indicate that serious injuries from TASERs are extremely rare. A 3-year review of all TASER uses against criminal suspects at six law enforcement agencies found only three significant injuries occurred out of 1,201 uses and also found that 99.75% of all criminal suspects shocked by a TASER received no injuries or only mild injuries such as scrapes and bruises.[18] Notwithstanding the apparent safety of CEDs, medical experts continue to recommend that any person exposed to CED activation be assessed for injuries and appropriate medical evaluation provided when nontrivial injuries are apparent or suspected. One reason for this recommendation continues to be that an existing medical or psychiatric condition in the suspect may have contributed to the behavior that led the police to use the CED to subdue the suspect, and that underlying condition may require medical assessment and treatment, independent of the CED exposure.[19]

Operational Success of CEDs

Several law enforcement agencies have reported good operational success from the use of CEDs. The Miami Police Department (Florida) reported that after it adopted CEDs in 2003, the department experienced no police-involved shootings for 20 consecutive months. During the same time period, the Seattle Police Department (Washington) also reported fewer police-involved shootings after adopting CEDs. Others, such as the Phoenix Police Department (Arizona), reported the lowest rate of police shootings in 14 years, along with the Madison Police Department (Wisconsin) that concluded that the deployment of CEDs reduced officers' use of deadly force and also reduced injuries to both officers and suspects.[20]

Scenario

On March 5, 2008, a bench warrant was issued for the arrest of 35-year-old Tim Jones for failure to appear in court to face several felony charges that had been filed against him because of his actions during an incident that occurred earlier. In that earlier incident, an officer of a large metropolitan police department used force to subdue Jones because of his reaction to his inquiries about his unusual behavior (walking the streets at 3:00 A.M. and making statements that he was "looking for the people that shot up his neighborhood"). Jones was described in a use of force report, filed by Officer D. Smith in the incident, as an emotionally disturbed person (EDP).

On April 19, 2008, at approximately 9:45 P.M., police officers J. Cushing and K. Dobbs went to Jones' residence to serve the bench warrant on Jones.

After arrival at Jones' residence, Cushing and Dobbs knocked on the door. Jones did not respond to the knocking even though they could see him inside the residence. However, Officer Cushing reported that Jones did say, "The police had killed his entire family."

Jones finally did respond to their knocking, at which time the officers noted that he had armed himself with a large knife and a frying pan. Officer Dobbs then requested a backup officer to come to the scene.

Shortly thereafter, Officer Harris (acting sergeant for the platoon) arrived and opened the sliding glass door to Jones' apartment. Jones immediately rushed toward the door armed with the knife and frying pan. Jones was ordered by the officers to drop the knife and frying pan, but he refused to do so. In response, Officer Dobbs fired multiple pepper-ball rounds at Jones in an unsuccessful attempt to disarm him. Several of the pepper-ball rounds struck Jones in the face.

At approximately 11:45 P.M., the special weapons and tactics (SWAT) team and the hostage negotiating team (HNT) were called to the scene.

At approximately 15 minutes after midnight (early morning of April 20, 2008), the HNT negotiator arrived on the scene and attempted, without success, to get Jones to talk with him.

At approximately 49 minutes after midnight, the SWAT team utilized a flash-bang device to distract Jones and simultaneously fired several beanbag sock rounds at him in an unsuccessful effort to disarm him. SWAT team member J. Johnson reported that "Jones jumped a little bit when the flash-bang went off," but up to that point, "he thought Jones could have been deaf" because of his failure to react to anything.

Shortly thereafter, SWAT sent in a K-9 dog in an effort to take Jones into custody. Officer Krinky, a K-9 handler, entered the apartment to retrieve the K-9 when Jones stabbed the K-9 dog. Jones then attempted to stab Krinky when he made his attempt to retrieve the dog. In an effort to defend Krinky, Police Sergeant Oliver Gumm shot Jones twice with his MP-5 rifle.

Although Jones was bleeding from his wounds, he continued to hold onto the knife and frying pan while standing near the kitchen sink staring at the officers.

At approximately 1:07 A.M., Major D. Brinkley, Commander of SWAT, deployed three canisters of tear gas (CS) in an unsuccessful effort to disarm Jones.

At approximately 1:29 A.M., SWAT made an unsuccessful effort to disarm Jones by spraying high-pressure water on him from a 2-inch fire hose.

At approximately 1:32 A.M., SWAT fired more beanbag rounds at Jones in an unsuccessful effort to disarm him.

Finally, Jones retreated to the bathroom and closed the door. It was then discovered that he had dropped the knife on the bedroom floor and was no longer armed. The officers then forced open the bathroom door and took Jones into custody. He was transported to a hospital where he died on June 21, 2008, as a result of the wounds he received during the incident.

How Would You Have Handled the above Scenario?

In 1987, the IACP and the U.S. Department of Justice, Bureau of Justice Assistance (BJA), established a National Law Enforcement Policy Center. In 1997, they published a Concept and Issues paper titled *Dealing with the Mentally Ill* and a Model Policy titled *Dealing with the Mentally Ill*. The purpose of these documents was to provide guidance to officers on how to respond when encountering people suspected as being mentally ill.

Section IV A of the Model Policy recommended that officers evaluate the behavior of people they come into contact with and make a judgment about the individual's mental state and the need for intervention absent the commission of a crime. One of the behaviors they recommend officers be alert for is the degree of reaction they get from the individual.

Section IV C of the Model Policy recommended that officers, when encountering mentally ill persons, take steps to calm the situation; assume a quiet, nonthreatening manner when conversing with the individual; move slowly so as to not excite the disturbed person; provide reassurance that the police are there to help; and communicate with the individual in an attempt to determine what is bothering him or her.

In this scenario, Jones gave no reaction at all to their knocking on the door and made no verbal comments except to say that the police had killed his entire family. That statement by Jones should have caused Officers Cushings and Dobbs to suspect that they were dealing with an emotionally disturbed person (EDP).

In this scenario, Jones was refusing to obey the officer's orders to drop the knife and frying pan. However, he was contained by the police in his apartment and was not a threat to anyone except himself. As long as the officers continued to contain and control the scene (prevent Jones from leaving and anybody else from entering), physical violence was not imminent.

Notwithstanding this being the situation, Officer Dobbs armed himself with a pepper-ball gun, opened the sliding glass door to the apartment, and ordered Jones to drop the knife. When Jones did not comply, he fired a minimum of 15 to 20 rounds all over Jones' body.

A pepper-ball is a projectile containing OC that is designed to break open upon impact, enveloping the individual in a cloud of OC. Pepper-ball devices are designed to give the police the capability to deliver OC from a safe distance which, upon getting in a person's eyes and breathing passages, causes temporary incapacitation due to pain, loss of vision, and difficulty in breathing. The theory behind its use is that during the period of temporary incapacitation, the police are able to take the suspect into custody. It is considered a less-than-lethal weapon if used in the manner and method intended.

However, the failure of Officers Cushing and Dobbs to recognize Jones was an EDP before employing force against him resulted in their use of a pepper-ball gun to fire pepper-balls that hit Jones in the face. That use of force, in the manner delivered, was potentially deadly.

Because pepper-balls have varying levels of kinetic energy, depending upon the distance from the suspect when fired, training protocol is that they not be fired to strike the head of a suspect. In October 2004, a college student in Boston, Massachusetts, died after being struck by a pepper-ball in the eye following a World Series baseball game. That incident demonstrated that a pepper-ball can have lethal results if not utilized in a manner consistent with training protocols and the pepper-ball manufacturer's recommendations.

The officer in command (OIC) in this scenario made a decision to engage Jones with SWAT before a trained negotiator could make a reasonable effort to get Jones to be responsive and before intelligence could be gathered which would have resulted in Jones being classified as an EDP. This decision to prematurely engage Jones with SWAT escalated the situation.

Finally, Officer Krinky, the K-9 handler, entered the apartment when Jones stabbed his K-9. Krinky should have been instructed that he was not to enter the apartment under any circumstances because his entry placed him in jeopardy and created the need to use deadly force against Jones to protect the life of Officer Krinky.

The failure of the officers involved in this scenario to follow the protocols for dealing with EDPs, coupled with their failure to follow protocols on the use of force (pepper-balls), coupled with a pattern of tactical errors (use of K-9 handler who was not instructed that he should not enter the apartment) and the continued escalation of the levels of force, when it should have been apparent that they were dealing with an EDP, resulted in the death of Jones.

Summary

Law enforcement can, in some incidents, reduce injuries to suspects and officers by using less-lethal weapons that balance minimal force with their operational need to arrest or disarm a dangerous person. The most reasonable force is the type that gets the job done and results in the least injury to suspects and officers.

Less-lethal weapons have proven to be effective tools that result in fewer injuries to suspects and officers than conventional police tactics. Therefore, less-lethal weapons should be used before conventional police tactics whenever the dynamics of the situation allow the officer to have a choice.[21]

Some less-lethal devices have generated civil litigation; however, in most instances, the courts have not deemed the use of less-lethal weapons to be unreasonable or unlawful per se, but have examined the lawfulness of the devices on a case-by-case basis. The courts have done this to determine whether, under the circumstances of the particular incident, the officers were justified in employing the device at all, and, if they were justified, whether the officers used the devices in the proper manner. Therefore, it is essential that officers use the devices correctly, in accordance with the manufacturer's recommendations, in accordance with department policy, and only to the extent necessary to accomplish the desired result.[22]

Endnotes

1. International Association of Chiefs of Police (IACP) Model Policy, *Less-Than-Lethal Weapons*, April 2002.
2. U.S. Department of Justice, Bureau of Justice Statistics, 2003.
3. *Conducted Energy Devices: Development of Standards for Consistency and Guidance*, Foreword by Chuck Wexler, Executive Director, PERF, November 2006.
4. Adam Bowles, *The American News Service*, Article No. 1621, October 26, 2000.
5. Ken Hubbs, *The Tactical Edge*, Spring 1997.
6. Vivian B. Lord, *Suicide By Cop: Inducing Officers to Shoot*, Looseleaf, New York, 2004.
7. *Plakas v. Drinski*, 19 F.3d 1143 (7th cir.) 1994.
8. D. Dorsch, *Law and Order*, September 2001.
9. Peter D. Button, *Less-Lethal Force Technology*, Metro Toronto Police.
10. New Jersey Attorney General's Model Policy on Use of Force, June 2000.
11. The Attribute-Based Evaluation of Less-Than-Lethal, Extended-Range, Impact Munitions, February 15, 2001.
12. Wound Ballistic Review, *Journal of the International Wound Ballistics Association* 4(4), Fall 2000.
13. Steven M. Edwards, John Granfield, and Jamie Onnen, *Evaluation of Pepper Spray*, National Institute of Justice, Research in Brief, February 1997.
14. John Granfield, James Onnen, and Charles Petty, *Pepper Spray and In-Custody Deaths*, IACP, March 1994, and Report for the U.S. Department of Justice by Charles S. Petty, February 2004.
15. *The Washington Post* (Washingtonpost.com), October 24, 2004, on the death of Victoria Snelgrove after being hit in the eye with OC pellet.
16. Conducted Energy Devices: Development of Standards for Consistency and Guidance, Introduction, Police Executive Research Forum and U.S. Department of Justice, November 2006.
17. Remarks of American Civil Liberties Union (ACLU) Legal Director Larry Dupuis in a letter to Madison, WI, police.
18. News Release, American College of Emergency Physicians, *Serious Injuries from TASER Are Extremely Rare*, January 15, 2009.

19. Ibid.
20. Conducted Energy Devices: Development of Standards for Consistency and Guidance, Police Executive Research Forum and U.S. Department of Justice, November 2006.
21. Greg Meyer, *Nonlethal Weapons vs. Conventional Police Tactics, the Los Angeles Police Department Experience.*
22. IACP National Law Enforcement Policy Center, *Policy Review* 13(2), 2001.

Officer-Involved Shootings (OISs)

3

Although, statistically, very few of the 800,000 law enforcement officers in the United States ever become involved in hostile shootings situations, 562 officers lost their lives during the time period from January 1999 through January 2009 from being shot.[1]

A prosecutor's office, when conducting a use of force analysis, seeks to determine whether or not there have been any violations of criminal statutes by the officer(s) involved (criminal culpability).

Police chiefs conduct their analyses to determine whether the involved officer(s) violated department rules, regulations, policies, protocols, or training (administrative culpability).

Attorneys representing the law enforcement agency and attorneys representing the person the force was used against conduct an analysis narrowly and specifically directed toward the determination of justification, or lack of justification, of the use of deadly force under the "objective reasonableness standard" established by the U.S. Supreme Court in the *Graham v. Connor*, 490 U.S. 386 (1989) case (civil culpability).

The seriousness of officer-involved shootings (OISs) cannot be overstated. The reputation and sometimes the careers of involved officers often depend upon whether a full and accurate determination can be made of the circumstances that precipitated the event and the manner in which it unfolded. The critical nature of these investigations is also underscored by the frequency with which these incidents result in civil litigation. From a broader perspective, a law enforcement agency's reputation within the community and the credibility of its personnel are also largely dependent upon the degree of professionalism and impartiality that the agency can bring to such investigations.[2]

Typically, detectives assigned to conduct an investigation into an OIS are experienced criminal investigators and are well qualified to conduct internal investigations into OISs; however, they sometimes tend to ignore evidence of wrongdoing on the part of their fellow officer and proceed from the assumption that the shooting was justified. As a consequence, all credibility of the agency is lost during a civil trial when arguments are successfully made by counsel for plaintiffs that the department, through the internal affairs investigation, ratified the conduct of the officers, and the custom and practice of the department is to ignore their officers' violations of constitutional rights. Police chiefs, sheriffs, and other law enforcement agency heads need to assign the best and most qualified investigators available to conduct the investigation into an OIS and instruct them to conduct a totally objective investigation.[3] The International Association of Chiefs of Police (IACP) National Law Enforcement Policy Center published a Concepts and Issues paper entitled "Investigation of Officer-Involved Shootings." A copy of that paper is provided in the appendix to this chapter, and law enforcement agencies are encouraged to read and implement the recommendations in that paper.

Law enforcement officers are armed because their duties cause them to go in harm's way, and they have a right to use deadly force in self-defense. When such an incident occurs, however, an OIS investigation should always follow. The investigation must be thorough, accurate, and timely, because the reputation and career of the involved officers can be negatively affected, and many OISs result in civil litigation. Failure to conduct a thorough, accurate, and timely investigation can result in the loss of important evidence, the filing of criminal charges that may not be warranted, the inappropriate assignment of culpability, and harm to the agency's reputation. For these reasons and others, certain hard questions should always be answered by investigators, including:

- What were the actual circumstances that precipitated the shooting, as described by participants and other witnesses?
- What does the physical evidence show, and how does it agree or disagree with the statements of witnesses?
- Did appropriate tactics precede the shooting, and if not, was that a factor?
- What did the shooting officer reasonably perceive, and when did he or she perceive it?
- Most importantly, was the shooting a reasonable exercise of self-defense?

Because of their unique role in our society, law enforcement officers who are involved in an OIS are generally accorded the benefit of any reasonable doubt and are rarely indicted on criminal charges of manslaughter or murder. Nevertheless, the investigation can be an extremely stressful experience for the officer involved. Even though police officers are trained to deal with traumatic and violent events, they can suffer from stress disorders following a shooting incident. Health professionals are more frequently finding that law enforcement officers involved in shooting incidents end up suffering some degree of posttraumatic stress disorder (PTSD). Accordingly, any OIS is to be avoided, if at all possible.[4]

What Would You Have Done?

This incident takes place in a large central city, with 2,700 municipal police officers and a population of mixed demographics. Basic firearms training includes computerized "shoot, don't shoot" simulations that are repeated annually, and the department's deadly force policy reflects the standards enunciated in both *Tennessee v. Garner*, 471 U.S. 1, (1985) and *Graham v. Connor*, 490 U.S. 386 (1989). The local media has been harshly critical of the two most recent shootings, wherein African American suspects were fatally shot under disputed fact patterns, but the district attorney ruled both of those cases to be justifiable self-defense.

Tyrone McGee was a 17-year-old African American youth who lived in a predominantly low-cost public housing project with his working mother and four younger siblings. He had not been in trouble and was not a member of any gang, but there were gang members in his neighborhood, and he knew that they were dangerous to anyone they encountered. He had obtained a driver's license, after successfully completing his high school driver education class, and his mother would occasionally let him drive her old car to run errands for her. On this particular hot summer evening, she asked Tyrone to go to the local convenience store for some cooking oil that she needed to prepare dinner, and eager to drive the car, as always, he quickly pulled on a T-shirt and left.

While he was en route to the store, the clerk called 911 to report that she had just been robbed at knifepoint and described the suspect as an African American, male teenager, wearing a white T-shirt, and driving a dark-colored, older-model, foreign car. Two young African American detectives, who worked neighborhood street crimes in rough clothing and an unmarked vehicle, monitored the radio broadcast of the robbery and the suspect's description. Because they were only three blocks away and the assigned patrol unit was farther out, they decided to roll on the call, too. Just as the detective riding shotgun exited their vehicle and was walking toward the store's entrance, Tyrone drove into the parking lot in his mother's 1992 black Toyota. Given the totality of circumstances, any experienced law enforcement officer would reasonably investigate this *suspect* and draw his or her weapon before making contact, which is exactly what the detective did. Tyrone, of course, knew nothing about the reported robbery and only saw a young African American man in rough clothing coming toward him with a gun. It would be disputed whether or not the detective displayed a badge or made verbal identification, but if he did, Tyrone neither saw nor heard, and his only thought was to get away immediately. In sheer panic, Tyrone floored the gas, and his mother's old car seemed to leap forward, as he turned for the street to flee. While making that turn, however, his headlights swept across the detective, who suddenly found himself at least momentarily in the vehicle's path. As he jumped to his left and the vehicle continued the turn to his right, he fired two quick shots at the driver and then two more as the vehicle continued on past. All four shots struck the vehicle, and one continued through the open passenger window to strike Tyrone in the head, killing him instantly, and his vehicle continued into the street to hit a passing car.

The OIS investigation quickly determined that Tyrone was not the robbery suspect and next had to address the detective's justification for shooting in what he asserted was self-defense to prevent being run over by the fleeing vehicle. As part of the aftermath, the media exhorted the community to demand *justice*, and the shooting detective was left to contemplate an ordeal that promised to be long and emotionally very stressful.

Best Practices

Any OIS is sure to be controversial under the best circumstances, and although the first shot may have been justified in self-defense of being run down by the decedent's vehicle, it is difficult to explain how the shooter was at risk when the final and fatal shot was fired from the side of the vehicle as it was passing by the shooter. This is not an unusual set of circumstances. The IACP has stated that

> Firearms shall not be discharged at a moving vehicle unless a person in the vehicle is immediately threatening the officer or another person with deadly force by means other than the vehicle. The moving vehicle itself shall not presumptively constitute a threat that justifies an officer's use of deadly force. An officer threatened by an oncoming vehicle shall move out of its path instead of discharging a firearm at it or any of its occupants.[5]

In any event, an OIS investigation should be conducted to the same standard of care as any other shooting, and it should be thoroughly reviewed for compliance with the department's policy and training. That essential review can also serve to make policy and training modifications that can better serve to prevent future events.

One of the most important areas in the investigation of police-related deaths is the crime scene investigation, as the collection and preservation of the relevant forensic evidence can prove or disprove issues and hypotheses. Below you will find an outline of appropriate crime scene procedures and a list of potential categories of forensic evidence that should be collected and sent to the laboratory for analysis.

Crime Scene Investigation

All the potential evidence, bullets, casings, and weapons should be secured in the original location and thoroughly documented before removal from the scene. One of the common mistakes made in shooting scenes is to dig out the

Figure 3.1 An overall view of a shooting death scene. (**See color insert following page 78.**)

bullets and pick up the evidence before the scene is documented. This type of failure will make the subsequent reconstruction impossible.

Each piece of potential evidence should be carefully studied, documented, collected, and preserved for further laboratory examination. Although an indoor scene is depicted in Figure 3.1, it graphically points out the extensiveness of the work to be accomplished during a crime scene investigation.

The crime scene investigation should be conducted by the agency's most experienced and best trained investigators. The location of bloodstains, bullets, casings, and weapons should be noted and documented. The position and location of the decedent also should be noted and documented.

The entire scene should be thoroughly documented using the standard crime scene documentation techniques of photography and sketches. The following areas are especially important to the successful reconstruction of the event:

- Photographs and measurement of bullet holes.
- Photographs and documentation of all patterns and wounds.
- Determination and documentation of bullet impact sites and their direction.

- Determination and documentation of gunshot residue (GSR) pattern evidence on the victim's body and clothing.
- Location and documentation of all spent bullets and casings.
- Location and documentation of all involved weapons.
- Photographs and documentation of all bloodstain patterns.

The Crime Scene Search

A combination of zone and link methods could be used for this type of scene search. Because this scene is limited to a confined area, the number of crime scene investigators allowed to enter the scene should be limited to avoid accidental contamination or alteration of the scene.

Collection, Preservation, and Packaging of Physical Evidence

The following types of forensic evidence must be collected in all OISs:

I. Evidence from Police Officer
   A. Weapons and ammunition
   B. Hand swabs for GSR
   C. Clothing and shoes
   D. Statements related to incident
II. Evidence from Decedent
   A. Clothing and shoes
   B. Blood patterns and GSR patterns are extremely important and should carefully preserved
   C. GSR hand swabs
   D. Gunshot wounds and other injury patterns
   E. Medical and autopsy reports
   F. Toxicology report
III. Evidence from Scene
   A. Bloodstains and their patterns
   B. Body position and location
   C. Guns, knives, and other weapons
   D. Bullets, casings, and fragments
   E. Bullet trajectory
   F. Ricochet and deflection patterns
   G. GSR and GSR patterns
   H. Markings, damages, and other patterns
   I. Fingerprints, shoe prints, and other imprint evidence
   J. Trace and transfer evidence such as hairs and fibers
   K. Alcohol or drug containers

Preliminary Reconstruction

One of the most important aspects of this investigation is to reconstruct the shooting event. Reconstruction is based on the forensic and pattern evidence found at the scene. High-velocity blood spatters and the bullet holes could be used to determine the location and position of the decedent. The bullet trajectory and spent casing locations are useful information to determine the officer's position and location. GSR patterns found on the body and clothing could used to estimate the distance between the weapons to the target. A more detailed reconstruction should be done after reviewing the autopsy results and forensic laboratory reports and evaluating witness observations and officer statements. This information should be useful to reconstruct the sequence of the event and to determine what, where, when, who, why, and how this shooting occurred. Reconstruction could also be used to confirm or refute the statements from witnesses, suspects, and police officers. Figure 3.2 provides a general scheme for the examination of forensic evidence.

Releasing the Scene

Scenes located inside of a house do not require urgency; however, scenes that are outside and subject to environmental changes should be processed as quickly as possible to prevent a risk of evidence contamination resulting with weather changes.

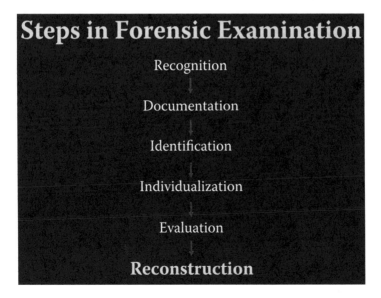

Figure 3.2 Steps in forensic examination of a crime scene. **(See color insert.)**

Laboratory Analysis and Reconstruction

Physical evidence collected from the crime scene should be submitted to a forensic science laboratory for further analysis and reconstruction. The following results may be obtained:

I. Ammunition
   A. Number of projectiles fired, number of projectiles found at the scene or in the body
   B. Projectile identification: bullet, shot pellet, bean pellets, slug
   C. Lands, caliber, gauge, shot size
   D. Class characteristics: manufacturer, style, type of weapon fired from
   E. Individual characteristics, striations, land-and-groove marks
   F. Damage and ricochet marks
   G. Trace evidence: blood, DNA, hair, tissues, fiber, wood, wall material

Cartridge Cases / Shells

- Number of spent casings or shells found at scene consistent with the number of shots fired. Each of the casings should be examined and identified
- Manufacturer
- Caliber and load specifications
- Factory versus reloads
- Individual characteristics
- Firing-pin impression
- Breech face marks
- Chamber marks
- Extractor/ejector marks
- Primer: center versus rim fire
- Powder: black powder versus smokeless
- Wad, cup, and other fillers

Weapon Examination

- Functionality—proper firing mechanism
- Firing pin, breech lock, ejector, extractor mechanism
- Loading mechanism
- Single, automatic, double barrel

- Magazine or cylinder
- Trigger pull
- Trace/fingerprints/DNA
- Ownership history

Gunshot Residues (GSR)

Gunshot residues consist of unburned powder particles, primer residues, lubricants, and barrel residues. These materials may be found either on target surfaces or on an individual's hand that discharged a firearm. The source of GSR is different:

- GSR from muzzle
  - Powder particles on target surface, such as skin, body, wound, clothing, object, surface
- GSR from cylinder gap
  - GSR particles on hands, clothing
- GSRs generally can be identified by atomic absorption (AA) or scanning electron microscopy with energy-dispersive x-ray (SEM-EDAX) analysis. GSR patterns are usually identified by the following techniques:
  - Visual examination
  - Microscopic examination
  - Infrared (IR) photography
  - Lead particles identification
- GSR particles identification

Examination of Ricochet Bullets

Each projectile should be examined for the evidence of ricochet. This finding is valuable for trajectory reconstruction. Projectiles may impact different types of materials or surfaces such as yielding surfaces of soil, sand, water, tissue, sheet metal, wood, or drywall; frangible surfaces, such as cinder blocks, bricks, or stepping stones; or nonyielding surfaces such as stone, concrete, steel, or heavy metal. When a projectile impacts on nonyielding surfaces, the following may occur:

- The projectile will lose some energy.
- The depart is usually different than the incident angle.
- The projectile usually has deformation.
- Trace transfer evidence will usually be present.
- The impact surface may play an important role.

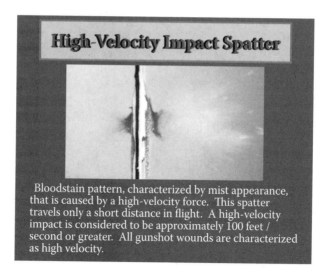

Figure 3.3 High-speed close-up photograph of high-velocity impact blood spatter. **(See color insert.)**

Bloodstain Pattern and Tissue Examination

Blood, tissues, and hairs often may be recovered at a shooting scene. The identification of these materials can provide valuable information for reconstruction. Figure 3.3 depicts a view of the production of high-velocity blood spatters from a gunshot.

The following should be checked for the presence or absence of biological material:

- Tissue, blood deposits on officer's clothing.
- Tissue, blood deposits on decedent's clothing.
- Tissue, blood deposits on weapons.
- Blood flows and drip patterns on body.
- High-velocity blood spatters on hands or wall.
- High- and medium-velocity impact blood spatters on arms.
- Bloodstains and aerial deposits on furniture.
- Blood, tissue deposits on floor or wall.
- DNA typing of those blood and tissue.

Figure 3.4 shows blood, hair, and tissues that were found on the barrel of the gun.

This information indicates that this gun was in direct contact with the entrance gunshot wound.

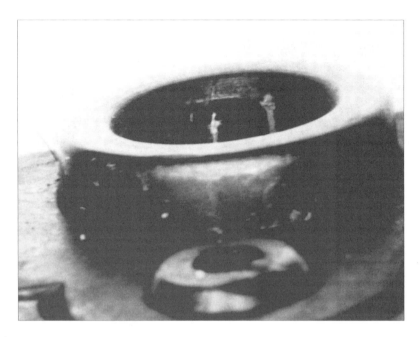

Figure 3.4 Close-up view of the biological evidence found on the gun barrel. **(See color insert.)**

Reconstruction

There are many types of reconstruction. In this particular case, the following types of reconstruction should be conducted:

- Wound pattern analysis
- Distance estimation
- Blood spatter pattern analysis
- GSR pattern determination
- Gunshot wound pattern analysis
- Trajectory determination
- Casing ejection pattern analysis
- Bullet damage and ricochet pattern
- Location and position of police officer
- Location and position of decedent
- The sequence of the shooting event

Figure 3.5 through Figure 3.8 depict different types of information generally obtained from reconstruction.

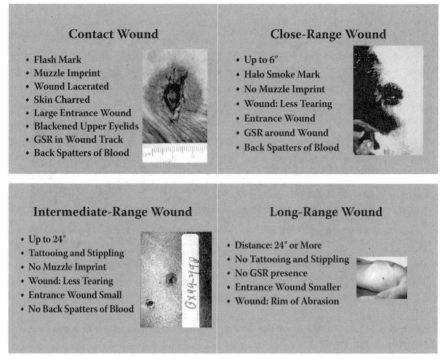

Figure 3.5 Gunshot wound pattern analysis. **(See color insert.)**

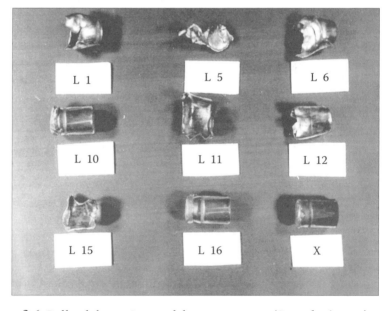

Figure 3.6 Bullet deformations and damage patterns. **(See color insert.)**

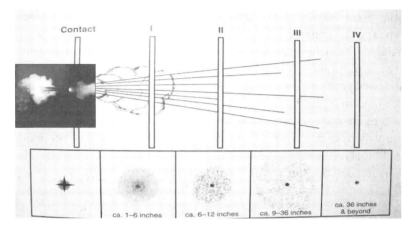

Figure 3.7 Gunshot residue pattern analyses. (**See color insert.**)

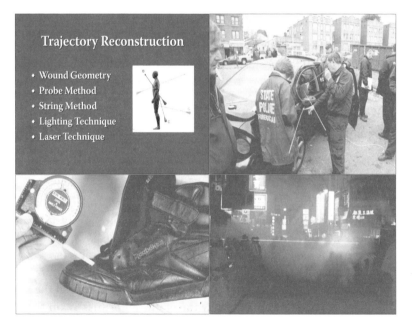

Figure 3.8 Techniques used for bullet trajectory reconstruction. (**See color insert.**)

Pathology / Medical Examination

External Examination

The body is of a well-developed, well-nourished, 5'6", 150-lb, African American man whose appearance is consistent with the given age of 17 years. The scalp hair is curly, black, and 1 to 1 1/2 in. or less. There is a 1/4-in.

mustache and slight, barely perceptible beard stubble. The irises are dark brown, and the conjunctivae are free of petechiae, jaundice, or hemorrhage. There is slight tache noire artifact on the left bulbar conjunctiva. The oral cavity has natural teeth in good condition. There are no injuries to the buccal mucosa, gingivae, or frenulae. The genitalia are of a normal, circumcised, adult man. Injury to the scrotum is described below.

There are no tattoos. The fingernails are markedly short with slight, focal accumulations of dirt. There are no injuries to the hands or left foot. Injuries to the right foot are described below. There is no edema of the extremities. There is a 1/4-in. nondescript scar on the anterolateral aspect of the right side of the neck, a 3/8-in. scar on the anterior aspect of the distal right forearm, a 1 1/2-in. curvilinear, slender scar on the anteromedial aspect of the distal left arm, a 3/8-in. circular scar on the lateral aspect of the left arm, and a 1/4-in. circular scar on the front of the right ankle. Two 1/4-in. scars are on the dorsum of the left hand.

Postmortem Changes

Rigor mortis is moderate and symmetrical in the extremities. Livor mortis is barely perceptible on the posterior portions of the body, and the body is cool.

Therapeutic Procedures

There are none.

Clothing

The body is received clad in gray undershorts, gray long underwear, and blue and black short-sleeved shirts. Received with the body are black pants, a brown belt, a blue and white coat with a hood, two dark shoes, and a red cap.

There is patchy blood staining on the inner lining of the blue and white coat and blood staining of the undershorts and long underwear, primarily on the left aspect. The short-sleeved shirts are blood stained and cut or torn in the front.

There are multiple perforations of the clothing, primarily on the lower aspects of the upper garments, and scattered about the lower garments. There is a perforation on the front of the upper aspect of the coat. There is no perforation on the upper back region of the coat. The red cap has two perforations toward the superior aspect. Both short-sleeved shirts have perforations on the front and back of the upper aspects of the shirts. There is a perforation on the front left aspect of the belt. There are several perforations of the right shoe, including a perforation on the medial aspect of the mid/distal shoe, a perforation on the dorsum of the mid/distal shoe, and a perforation on the plantar surface of the distal shoe.

The clothing and bags from the hands are retained.

Injuries: External and Internal

There are multiple (19) gunshot wounds to the body, including a perforating gunshot wound to the anterior chest, a perforating gunshot wound to the mid left trunk, perforating gunshot wounds (4) to the left side of the lower trunk, a perforating gunshot wound to the left back, a perforating gunshot wound to the right elbow, perforating gunshot wounds (2) to the lateral aspect of the left thigh, perforating (1) and penetrating (1) gunshot wounds to the posterolateral aspect of the left thigh, perforating gunshot wounds (2) to the left lower extremity, and perforating (3) and penetrating (2) gunshot wounds to the right lower extremity. The wounds are lettered for descriptive purposes only; no sequence is implied. Directions of travel are stated with the body in the normal anatomic position.

Perforating Gunshot Wound to Anterior Chest

There is a perforating gunshot wound (A) to the mid anterior chest above the level of the nipples, 15 in. below the top of the head and 3/4 in. to the left of midline. The 1/4-in. circular/oval perforation has a 1/8-in. margin of abrasion which is most prominent on the inferior aspect. There is no fouling or stippling of the adjacent skin.

After perforating the skin of the anterior chest, the bullet perforated the left second intercostal space, grazing the lateral edge of the sternum. Several small pieces of bone are on the inner aspect of the sternal perforation. The bullet continued from front to back, perforating the arch of the aorta and causing a 1 1/2-in. defect at the junction of the left subclavian and left common carotid arteries. There is marked mediastinal hemorrhage. The bullet perforated the medial aspect of the upper lobe of the left lung prior to perforating the left aspect of the thoracic vertebra and spinal cord, causing marked softening of the thoracic cord.

There is patchy epidural and subdural hemorrhage surrounding the spinal cord. The bullet exited the posterior aspect of the third thoracic vertebra before exiting the mid upper back. In the left pleural cavity are 1250 cc of liquid blood, and 1000 cc are in the right pleural cavity. The bullet did not enter the pericardial sac; however, there is moderate hemorrhage on the superior aspect of the pericardium. There are no injuries to the heart.

The exit wound on the upper back is 11 in. below the top of the head and 3/4 in. to the left of midline. The 1/8-in. irregular perforation has no margin of abrasion or microtears of the margin. The direction of travel is front to back and upward. No bullets or fragments are recovered.

Perforating Gunshot Wound to Mid-Left Trunk

There is a perforating gunshot wound (B) to the lateral aspect of the mid left trunk, 21 1/2 in. below the top of the head. The 1.4-in. circular/oval perforation has a 1/8-in. margin of abrasion. There is no fouling or stippling of the adjacent skin.

After perforating the skin, the bullet perforated the left tenth intercostal space and briefly entered the inferior aspect of the left pleural cavity without perforating the lung.

The bullet then perforated the diaphragm and spleen, causing a 2-in. irregular perforation in the spleen. The bullet grazed the superior aspect of the left kidney causing slight, barely perceptible subcapsular hemorrhage on the kidney. The bullet then entered the peritoneal cavity, perforating several loops of small intestine and causing barely perceptible or slight hemorrhage adjacent to the perforations. There is slight perinephric hemorrhage. There are 250 cc of liquid blood in the peritoneal cavity.

The exit wound on the lateral aspect of the right side of the abdomen and at the level of the umbilicus is 25 1/2 in. below the top of the head. The 3/8-in. slit-like perforation has no margin of abrasion. The direction of travel is left to right with slight back to front and slight downward deviation. No bullets or fragments are recovered.

Perforating Gunshot Wounds (4) to Left Aspect of the Lower Trunk

There are four gunshot wounds (C, D, E, F) to the left aspect of the lower trunk, three of which (D, E, F) are clustered together. Due to their close proximity, intersection of the wound tracks, and paucity or absence of associated hemorrhage, the tracks (D, E, F) cannot be separated from each other. Hence, they will be described together.

There is a perforating gunshot wound (C) to the left anterolateral aspect of the lower trunk, 27 in. below the top of the head. The 3/16-in. circular perforation has a 1/16-in. to 1/8-in. margin of abrasion which is greatest along the anterior aspect. There is no fouling or stippling of the adjacent skin.

After perforating the skin, the bullet perforated the left iliac crest as it traveled upward and front to back. The bullet did not enter the peritoneal cavity. There is slight to moderate hemorrhage on the mid left back associated with the wound track.

The exit wound on the mid left back is 21 1/2 in. below the top of the head and 1 1/2 in. to the left of midline. The 1/2-in. irregular perforation has no margin of abrasion. The direction of travel is left to right, upward, and front to back. No bullets or fragments are recovered.

There are three clustered perforating gunshot wounds (D, E, F) on the lateral aspect of the left hip 29 to 30 in. below the top of the head. The 1/4-in. circular perforations have 1/16- to 1/8-in. abraded margins. Gunshot wound (D) has a margin of abrasion that is greatest on the posterior aspect. Gunshot wounds (E) and (F) have circumferential margins of abrasion.

There is no fouling or stippling of the adjacent skin. Injuries associated with these wounds are perforations of the left pelvis and intestines, and slight to moderate intramuscular hemorrhage within the lower back. There is an exit wound on the mid/lower right back 22 in. below the top of the head and 2 1/4 in. to the right of midline.

The 3/4-in. slit-like/irregular perforation has no margin of abrasion. There is an exit wound on the left lower abdomen just above the inguinal crease, 29 in. below the top of the head.

The 1 1/16-in. oval perforation has no margin of abrasion; however, there is a 1 1/4-in. × 3/8-in. abrasion adjacent to the superomedial aspect of the perforation. There is an exit perforation on the anterolateral aspect of the proximal right thigh, 34 1/2 in. below the top of the head. The 1/2-in. slit-like perforation has no margin of abrasion. There are no perforations of the bladder, rectum, or prostate; however, there is slight to moderate pelvic soft tissue hemorrhage associated with these wounds. The directions of travel of gunshot wounds (D), (E), and (F) are left to right. No bullets or fragments are recovered.

Perforating Gunshot Wound to Left Back

There is a perforating gunshot wound (G) to the mid left back, 21 in. below the top of the head and 3 in. to the left of midline. The 1/8-in. circular perforation has a 1/16-in. margin of abrasion on the inferolateral aspect. There is no fouling or stippling of the adjacent skin.

After perforating the skin, the bullet perforated the tenth thoracic vertebra and spinal cord causing barely perceptible hemorrhage adjacent to the lower thoracic cord. There is slight to moderate hemorrhage within the mid left back. The bullet then entered the right pleural cavity, sequentially perforating the medial aspect of the lower lobe of the right lung, the diaphragm, the superior aspect (dome) of the right lobe of the liver (causing a 3-in. defect), the diaphragm, and reentered the pleural cavity, perforating the lateral aspect of the lower lobe of the right lung. There is focal hemorrhage adjacent to the lung perforations. The bullet then exited the pleural cavity by perforating the lateral aspect of the right eighth intercostal space and grazing the superior aspect of the right ninth rib. The exit wound is on the lateral aspect of the right trunk, 16 1/2 in. below the top of the head. The 3/16-in. slit-like, oval, nondescript perforation has no margin of abrasion. The direction of travel is left to right, upward, and slightly back to front. No bullets or fragments are recovered.

Perforating Gunshot Wound to Right Elbow

There is a perforating gunshot wound (H) to the medial aspect of the right elbow, 21 1/2 in. below the top of the head. The 1/8-in. circular perforation has a 1/8-in. margin of abrasion. There is no fouling or stippling of the adjacent skin. After perforating the skin, the bullet perforated the distal right humerus before exiting the posterolateral aspect of the distal right arm. There is slight, focal hemorrhage adjacent to the wound track.

The exit wound on the posterolateral aspect of the distal right arm is a 5/8-in. irregular, nondescript perforation. The direction of travel is left to right, upward, and slightly front to back. No bullets or fragments are recovered.

Perforating Gunshot Wounds (2) to Lateral Aspect of Left Thigh

Gunshot wound (I) is a perforating gunshot wound to the lateral aspect of the proximal left thigh, 35 in. below the top of the head. The 1/4-in. circular/oval perforation has a 1/8-in. margin of abrasion that is greatest on the posterior aspect of the perforation. There is no fouling or stippling of the adjacent skin.

After perforating the skin, the bullet perforated the left thigh, causing barely perceptible or no visible hemorrhage along the wound track. The exit wound is on the anteromedial aspect of the proximal left thigh, 36 1/2 in. below the top of the head. The 5/8-in. oval/irregular perforation has no margin of abrasion. The direction of travel is left to right with slight back to front and barely perceptible downward deviation. No bullets or fragments are recovered.

Gunshot wound (J) is a perforating gunshot wound to the lateral aspect of the proximal left thigh, 36 1/2 in. below the top of the head. The ¼-in. circular perforation has a barely perceptible circumferential margin of abrasion and an irregular 1/2-in. nondescript abrasion adjacent to the inferior aspect. There is no fouling or stippling of the adjacent skin.

After perforating the skin, the bullet perforated the left thigh, causing barely perceptible or no visible hemorrhage along the wound track. The exit wound on the anterior aspect of the proximal left thigh is 37 1/2 in. below the top of the head. The 3/4-in. oval perforation has no margin of abrasion.

The direction of travel is left to right, slightly back to front, and barely perceptibly downward. No bullets or fragments are recovered. There are no injuries to the femoral vessels associated with these gunshot wounds.

Penetrating (1) and Perforating (1) Gunshot Wounds to Posterolateral Aspect of Left Thigh

Gunshot wound (K) is a penetrating gunshot wound to the posterolateral aspect of the proximal left thigh, 36 1/2 in. below the top of the head. The 7/8-in. oval perforation has a 1/16- to 1/8-in. margin of abrasion. There is no fouling or stippling of the adjacent skin.

After perforating the skin, the bullet perforated the gluteus muscles and entered the pelvic/peritoneal cavity. The wound may have contributed to perforations of the intestine. There is barely perceptible or no visible hemorrhage adjacent to the wound track. The direction of travel is left to right, upward, and back to front. A medium-caliber, slightly deformed, jacketed bullet is recovered free in the right lower quadrant of the peritoneal cavity. It is labeled "5" and retained.

Gunshot wound (L) is a perforating gunshot wound to the posterolateral aspect of the proximal left thigh, 37 1/4 in. below the top of the head. The 1/4-in. oval perforation has a barely perceptible, irregular margin of abrasion that is more prominent on the inferolateral aspect of the perforation. There is no fouling or stippling of the adjacent skin.

After perforating the skin, the bullet perforated the left thigh, causing barely perceptible or no visible hemorrhage along the wound track. The bullet exited the left side of the groin (at the junction of the scrotum and thigh) where there is a 1-in. nondescript perforation with no margin of abrasion. The bullet then caused a superficial perforation in the scrotum.

The superficial perforation in the scrotum has no margin of abrasion and no visible hemorrhage. There is no discernible wound track associated with the scrotal perforation.

The direction of travel is left to right, upward, and back to front. No bullets or fragments are recovered from the wound track.

Perforating Gunshot Wounds (2) to Left Lower Extremity

Gunshot wound (M) is a perforating gunshot wound to the anterolateral aspect of the distal left thigh, 45 1/2 in. below the top of the head. The 5/8-in. nondescript perforation has a 1/8- to 1/4-in. margin of abrasion that is slightly more prominent on the superolateral aspect of the perforation. There is no fouling or stippling of the adjacent skin.

After perforating the skin, the bullet passed a short distance across the distal left thigh before exiting the anterior aspect of the distal left thigh. There is no visible hemorrhage along the wound track.

The exit wound on the anterior aspect of the distal left thigh, just above the knee is 45 1/2 in. below the top of the head. The 3/8-in. perforation has no margin of abrasion. The direction of travel is left to right. No bullets or fragments are recovered.

Gunshot wound (N) is a perforating gunshot wound to the mid left shin, 55 1/2 in. below the top of the head. The 1/4-in. perforation has a barely perceptible margin of abrasion. There is no fouling or stippling of the adjacent skin.

After perforating the skin, the bullet perforated the left tibia and fibula before exiting the posteromedial aspect of the proximal left leg. A portion of the bullet exited the mid left shin 53 in. below the top of the head, where there is a 1/2-in. irregular perforation with no margin of abrasion. There is no visible hemorrhage along the wound track or adjacent to the perforated tibia and fibula.

The exit wound on the posteromedial aspect of the proximal left leg is 48 1/4 in. below the top of the head. The 3/4-in. slit-like perforation has no margin of abrasion.

The direction of travel is upward, front to back, and slightly left to right.

No bullets or fragments are recovered from the wound track.

Perforating (Gunshot Wounds O, P, R) and Penetrating (Gunshot Wounds Q and S) Gunshot Wounds to Right Lower Extremity

Gunshot wound (O) is a perforating gunshot wound to the anteromedial aspect of the mid/distal right thigh, 42 in. below the top of the head. The 1/2-in. irregular, nondescript perforation has a barely perceptible margin of abrasion that is more prominent on the inferior aspect. There is no fouling or stippling of the adjacent skin.

After perforating the skin, the bullet perforated the right thigh before exiting the anterolateral aspect of the mid right thigh. There is no visible hemorrhage associated with the wound track. The exit wound on the anterolateral aspect of the mid right thigh is 40 in. below the top of the head. The 1/2 in. nondescript perforation has no margin of abrasion. The direction of travel is left to right and slightly upward. No bullets or fragments are recovered.

Gunshot wound (P) is a perforating gunshot wound to the right ankle. The entrance wound is just medial to the Achilles tendon 4 in. above the base of the heel and consists of a 1/8-in. circular perforation with a 1/16-in. circumferential margin of abrasion. There is no fouling or stippling of the adjacent skin.

After perforating the skin of the heel, the bullet passed a short distance from left to right and exited the lateral aspect of the heel just behind and above the lateral malleolus, 4 1/2 in. above the base of the heel. The 3/8-in. irregular abrasion has no margin of abrasion. There is no visible hemorrhage adjacent to the wound track. The direction of travel is left to right, slightly upward, and slightly back to front. No bullets or fragments are recovered.

Gunshot wound (Q) is a penetrating gunshot wound to the medial aspect of the mid/distal right foot. The entrance perforation is a 5/8-in. nondescript perforation with a barely perceptible, irregular margin of abrasion. There is no fouling or stippling of the adjacent skin.

After perforating the skin of the foot, the bullet passed a short distance before lodging in the plantar aspect of the mid foot. There is slight to moderate hemorrhage associated with the short wound track. The direction of travel is left to right and slightly front to back. A medium-caliber, deformed jacket and bullet (two separate pieces) are recovered from the right foot. The number "6" is placed on the lead portion of the bullet, and both pieces are retained.

Gunshot wound (R) is a perforating gunshot wound of the right third toe. The similar appearing entrance and exit wounds are 1/4- to 3/8-in. irregular perforations, one adjacent to the front of the nail, and one on the dorsal surface of the mid toe. (Note that the direction of travel cannot be ascertained with certainty; however, it is likely slightly upward and slightly front to back relative to the normal anatomic position.) There is no fouling or stippling of the skin adjacent to either perforation.

Gunshot wound (S) is a penetrating gunshot wound to the lateral aspect of the distal right leg, 7 1/2 in. above the base of the heel. The 1/2-in. oval/irregular perforation has a 3/4-in. triangular abrasion on the inferior aspect of the perforation. There is no fouling or stippling of the adjacent skin.

After perforating the skin, the bullet passed upward through the right leg perforating subcutaneous and intramuscular tissue before lodging close to the right popliteal region at the posterolateral aspect of the proximal right leg. There is barely perceptible or no visible hemorrhage along the wound track, and focal hemorrhage adjacent to the entrance perforation and site of lodgment. The direction of travel is upward and slightly front to back. A medium-caliber, intact, jacketed bullet recovered from the posterolateral aspect of the proximal right leg is labeled "7" and retained. There is a 1-in. abrasion on the left lower abdomen.

These injuries, having been described, will not be repeated.

Internal Examination

Body Cavities

The organs are in the usual situs. The pleural and peritoneal blood accumulations were noted above. There is no pericardial accumulation. The surfaces are smooth and glistening.

Head

The scalp is atraumatic. There are no skull fractures, and there is no epidural, subdural, or subarachnoid hemorrhage. The 1400-gm brain is symmetrical and has normal gyri and sulci. The leptomeninges are smooth, delicate, and transparent, and the leptomeningeal vessels are normal. The arteries at the base of the brain are free of atherosclerosis. The cranial nerves have normal distributions. The surfaces of the brain stem and cerebellum are unremarkable. The cortical gray matter, subcortical and deep white matter, deep gray nuclei and ventricles are normal. The cerebrospinal fluid is clear. Horizontal sections of the brain stem and cerebellum are unremarkable.

Neck

The cervical vertebrae, hyoid, tracheal and laryngeal cartilages, and paratracheal soft tissues are normal. The upper airway is not obstructed.

A posterior neck dissection is unremarkable.

Cardiovascular System

The aorta and branches are free of atherosclerosis. The venae cavae and pulmonary arteries have no thrombus or embolus.

The 250-gm heart has a normal distribution of patent, right dominant coronary arteries. The myocardium is uniformly red brown without hemorrhage, softening, or pallor. The left ventricle wall is 1.3 cm. The endocardial surfaces, heart valves, chordae tendineae, and papillary muscles are normal.

Respiratory System

The right lung weighs 225 gm, and the left lung, 150 gm.

The lungs are moderately atelectatic, and visceral pleural surfaces are smooth and glistening. The bronchi are not obstructed. The vessels have no thrombus or embolus. The parenchyma is soft and spongy without consolidation.

Liver, Gallbladder, Pancreas

The 1025-gm liver has a smooth capsule. The parenchyma is diffusely red brown, moderately pale, and wet; the bile ducts are unremarkable. The small and moderately collapsed gallbladder contains several cc of dark green viscid bile without stones. The pancreas is uniformly tan/gray and has a normal lobular appearance.

Hemic and Lymphatic Systems

The 150-gm spleen has a smooth capsule. The parenchyma is firm with slightly distinct follicles. There are no lymph node enlargements. The thymus is atrophic. The bone marrow of the ribs and clavicles is unremarkable.

Genitourinary System

The kidneys weigh 100 gm each. The cortices are smooth and markedly pale. The vessels are free of atherosclerosis. The calyces and pelves are empty, opening into ureters that maintain uniform caliber and open into an unremarkable bladder containing approximately 50 cc of slightly red-tinged urine.

The prostate is unremarkable. There is barely perceptible hemorrhage within the right testis and within the epididymi associated with gunshot wound (L) noted above; however, there is no visible testicular perforation.

Endocrine System

The pituitary is unremarkable. The adrenals are normal. The thyroid is markedly pale, small, and symmetrical.

Digestive System

The esophagus is unremarkable. The stomach contains 200 cc of beige/green liquid/semisolid material with moderate, nondescript, partially digested food, including apparent pieces of carrots and green vegetable. The small and large intestines and appendix are unremarkable, except where noted previously.

Musculoskeletal System

The musculature is well developed and normally distributed. Perforations of the skeleton were noted above.

Additional Studies

Radiographs and photographs are obtained.

Examination of the anterior and posterior aspects of the extremities, the posterior neck, back, and buttocks are unremarkable, except where noted above in association with the gunshot wound injuries.

Gunpowder residue swabs of the palmar and dorsal surfaces of both hands are obtained.

Bullet Evidence

Three bullets recovered from the body were noted above. Five additional bullets are recovered from the clothing and gurney as follows: A medium-caliber, slightly deformed, jacketed bullet from the posterior aspect of the undershorts is labeled "X" and retained; a medium-caliber, intact, jacketed bullet from the posterior undershorts is labeled "4" and retained; three fragments from the long underwear including a jacket are retained with the number "3" placed on the base portion; a medium-caliber, slightly deformed, jacketed bullet from the long underwear adjacent to the ankles is labeled "2" and retained; a medium-caliber, slightly deformed, jacketed bullet from the gurney is labeled "1" and retained.

General Evidence

A small black and silver circular, foreign object is recovered from the gurney and retained. Multiple small and minute pieces of foreign material from the small right pant change pocket are retained.

Final Diagnoses

   I. Perforating gunshot wound to anterior chest (gunshot wound A), with:
  - A. Perforations of sternum, aorta, lung, third thoracic vertebra, spinal cord
  - B. Marked mediastinal hemorrhage
  - C. Hemothoraces (left equal 1250 cc, right equal 1000 cc)

  II. Perforating gunshot wound to mid left trunk (gunshot wound B), with:
  - A. Perforations of spleen, left kidney, and intestine
  - B. Hemoperitoneum (250 cc)

 III. Perforating gunshot wounds to left pelvic region/lower trunk (gunshot wounds to C, D, E, F), with:
  - A. Perforations of pelvis and intestines

 IV. Perforating gunshot wound to left back (gunshot wound G), with:
  - A. Perforations of tenth thoracic vertebra, spinal cord, right lung, liver, and right ninth rib

  V. Perforating gunshot wound to right elbow (gunshot wound H), with:
  - A. Perforation of distal right humerus

 VI. Perforating gunshot wounds to lateral aspect of left thigh (gunshot wounds I and J)

VII. Penetrating (gunshot wound K) and perforating (gunshot wound l); gunshot wounds to posterolateral aspect of left thigh, with:
  - A. Perforations of intestine and scrotum

VIII. Perforating gunshot wounds to left lower extremity (gunshot wounds M and N), with:
  - A. Perforations of tibia and fibula

 IX. Perforating (gunshot wounds O, P, and R) and penetrating (gunshot wounds Q and S) gunshot wounds to right lower extremity

  X. Abrasion, left lower abdomen

Cause of Death

Cause of death is multiple gunshot wounds to trunk with perforations of aorta, spinal cord, lungs, liver, spleen, kidney, and intestines.

Manner of Death

Manner of death is homicide (shot by police officers).

Endnotes

1. National Law Enforcement Officers Memorial Fund—Facts and Figures, *Causes of Law Enforcement Deaths* (1999–2008).

2. International Association of Chiefs of Police (IACP), *Investigation of Officer Involved Shootings*, Concepts and Issues Paper, 1999.

3. Ibid.

4. IACP, *Post-Shooting Incident Procedures*, Concepts and Issues Paper, March 1, 1991.

5. IACP, *Use of Force,* Model Policy 2-06.

Appendix: IACP National Law Enforcement Policy Center Concepts and Issues Paper Titled "Investigation of Officer-Involved Shootings" (August 1999)

I. Introduction

  A. Purpose of Document

This paper was designed to accompany the Model Policy on Investigation of Officer-Involved Shootings established by the IACP National Law Enforcement Policy Center. This paper provides essential background material and supporting documentation to provide greater understanding of the developmental philosophy and implementation requirements for the model policy. This material will be of value to law enforcement executives in their efforts to tailor the model to the requirements and circumstances of their community and their law enforcement agency.

  B. Background

Statistically, few officers become involved in hostile shooting situations. But all officers should have an understanding of steps that must be taken in such an event. The initial response of involved officers and the steps taken thereafter by first responders, supervisory, and investigative personnel often determine whether an accurate and complete investigation can be conducted. The accuracy and professionalism of such investigations can have significant impact on involved officers.

Other than in training exercises or similar agency-authorized actions, discharges of firearms by police officers, whether on or off duty, should be the subject of departmental investigation. The extent of the investigation should depend largely upon the real or potential impact of the shooting. Shootings that take place under hostile circumstances and, in particular, those in which injuries or fatalities have occurred, are situations that require more intensive investigation and involve a broader range of potential information requirements. The present discussion is focused primarily on this latter type of hostile-shooting investigation.

However, the discussion is also suitable as guidance for other shooting investigations.

The seriousness of officer-involved shootings cannot be overstated. The reputation and the careers of involved officers often depend upon whether a full and accurate determination can be made of the circumstances that precipitated the event and the manner in which it unfolded. The critical nature of these investigations is also underscored by the frequency with which these incidents result in civil litigation.

From a broader perspective, a law enforcement agency's reputation within the community and the credibility of its personnel are also largely dependent upon the degree of professionalism and impartiality that the agency can bring to such investigations. Superficial or cursory investigations of officer-involved shootings in general and particularly in instances where citizens are wounded or killed can have a devastating impact on the professional integrity and credibility of an entire law enforcement agency. An accurate and complete investigation of these deadly force incidents requires agency planning and the establishment of protocols that must be followed in such instances. It also depends largely upon the prudence of decisions made and steps taken immediately following shootings by the officers involved, supervisory personnel, and criminal investigators. Failure to take appropriate measures can lead to the loss of indispensable evidence, inaccurate investigative findings, inappropriate assignment of responsibility or culpability for wrongdoing, and even the filing of criminal charges against involved officers.

Many agencies, because of their limited resources and expertise in these matters, may rely in all or in part on the investigative resources and expertise of a state police agency, sheriff's department, or other law enforcement authority with appropriate jurisdiction. But these investigations cannot be turned over completely to others. For example, most of the burden for evidence preservation and protection of the crime and/or incident scene is the responsibility of involved officers and first responders. Therefore, it is essential that all officers have an understanding of the significance and importance of the proper initial officer responses and appropriate investigative measures required to conduct a professional officer-involved shooting investigation.

II. Procedures

    A. Involved Officer Responsibilities

        As indicated in the model policy, for officers involved in a hostile-shooting situation, there are four general areas of concern that should be addressed after the initial confrontation has been quelled: (1) the welfare of officers and others at the scene, (2) apprehension of suspects, (3) preservation of evidence, and (4) identification of witnesses.

The safety and well-being of the officer(s) and any innocent bystanders is the first priority. Initially, the officer should ensure that the threat from the suspect has been terminated. This includes but is not limited to handcuffing or otherwise securing the suspect. Should firearms or other weapons be available to or in the vicinity of the suspect, they should be confiscated and secured. All suspects should be handcuffed unless emergency life saving activities being employed at the time would be hindered by these actions. If not handcuffed or otherwise secured during the application of emergency first aid, an unencumbered armed officer must be present at all times to oversee security of the suspect and safety of emergency service providers. One should never assume that because a suspect has been shot or otherwise incapacitated that he is unable to take aggressive action.

The agency's communication center should also be provided with information on any suspects or suspicious persons who may have left the area, to include their physical description, mode and direction of travel, and whether the suspects are armed. A decision to pursue suspects will generally be based on a wide array of factors. However, the most important of these generally involves the ability of officer(s) at the scene to conduct a pursuit, the potential for apprehension of suspects, and the need to provide assistance to injured parties at the scene.

Given the diminished physical and mental condition of many officers following a shooting incident, it is better for these officers to stay at the scene. In so doing, they may be able to assist the injured, protect evidence, identify witnesses, provide dispatch with suspect information, and assist in establishing a containment area to aid in the apprehension of escaping suspects. Obviously, if the officer is injured, he/she should request emergency medical assistance as soon as possible. But, in the interim, the officer should administer emergency first aid to himself/herself to the degree possible.[1] Where reasonably possible, officers should then administer first aid to other injured parties pending the arrival of emergency medical service providers.

Assuming that one or more of the involved officers is physically capable of taking action following the incident, there are several other concerns that should be addressed immediately. For example, officers should request the presence of a supervisory officer as well as required back-up. Depending upon the circumstances, back-up assistance may involve a number of the agency's specialized units in addition to patrol units for traffic control, protection of evidence, and related matters. These may include command-level officers, a hostage negotiator, SWAT, K-9, crime scene technicians and a public information officer among other possibilities.

Immediately following hostile-shooting incidents, many officers are emotionally and physically disoriented. The ability of an officer to recognize and understand these problems is important in the officer's efforts to regain a degree of control over the situation and take appropriate measures. But this is often a difficult undertaking. Officers who have been involved in shootings often experience a number of immediate and involuntary physical and emotional reactions that may interfere with their ability to react effectively.

Emotional and physical reactions vary according to many factors involved in the shooting incident. These include the officer's perceived vulnerability during the incident, the amount of control he/she had over the situation and his/her ability to react effectively. It also includes such factors as how close the officer was to the victim, how bloody the incident may have been, and the nature or character of the suspect. In this latter issue for example, an incident involving a hardened and notorious killer can have a different effect on an officer than one which involved a scared teenager. Similarly, shooting a person who used the officer to commit suicide may evoke an angry response while other situations may produce far different feelings.

A variety of traumatic reactions caused by a shooting incident may interfere with an officer's ability to cope and react effectively and appropriately. For example, it is quite common for an officer involved in a hostile-shooting incident to experience perceptual distortions of various types. Some may experience time distortion in which events appear to occur in slow motion. For others, time may seem to accelerate. Auditory distortions are also common among officers involved in shootings. For most, sound diminishes and gunshots, shouts, or other sounds may be muffled or unheard. An officer may not hear all the shots being fired and may not be able to relate this type of information accurately if questioned at a later date. Involved officers should be aware of the possibility that their recall is impaired in one or more of these ways and investigative officers should keep these and related factors in mind when conducting shooting investigations.

For example, it is not advisable to conduct in-depth investigative interviews with officers immediately following their involvement in a shooting if they are experiencing such reactions. By the same token, in many situations, officers cannot provide basic or detailed information concerning the shooting. Officers should not be self critical because of this nor should investigators assume that this is necessarily an indication that the officers are purposefully withholding information. It is reasonable to entertain the possibility that these lapses are the result of trauma associated with the incident. Interviews conducted at a later time after the officer has had the opportunity to regain his/her composure may be more productive.

A complete discussion of the symptoms and effects of post-traumatic stress is beyond the scope and purpose of this paper. However, for purpose of the present discussion, it may be sufficient to recognize that such emotional and psychological phenomena are relatively common. As such, officers should be aware of these possibilities and recognize first, that they are natural responses to traumatic and unusual events and second, that the officers are not "going crazy" or responding in bizarre and irrational ways. In fact, they are exhibiting natural adaptive reactions to highly unusual life threatening situations.

Given the above context, it is often unrealistic for involved officers to perform many of the first aid and post-shooting actions that have been identified thus far and many of those that will be discussed. But officers must attempt to muster as much self-composure as possible in order to protect themselves and be cognizant of events around them. Understanding that they are experiencing one or more of these emotional or psychological reactions may assist officers in their attempts to regain their composure immediately following a shooting incident.

For example, officers need to be aware of their surroundings following a shooting and take note of important facts such as the time of day, lighting conditions, persons present, those who may have departed the scene, witnesses or potential witnesses, possible suspects or accomplices, and suspect vehicles, to name only a few. In some cases, emergency medical personnel and/ or fire fighters may be on hand prior to the arrival of back-up police personnel. Officers at the scene should make note of this with the understanding that these personnel may unknowingly move, misplace, or even inadvertently destroy evidence in the course of performing their duties. For example, it is not uncommon for such persons to remove (even ultimately discard) items of clothing from the suspect or others who have been wounded. Such items of clothing are often invaluable in attempts to establish the position, location, and distance of involved individuals in a shooting exchange.

Other items of potential evidentiary value should be of particular concern. One of the principal evidentiary items among these is firearms. In this regard, officers should ensure that their firearm is secured safely and that it is not handled in any manner until it can be examined by investigators or other designated police personnel. The firearm should not be removed if it is holstered. Nor should it be opened, reloaded, or tampered with in any other manner. In some instances the officer's or suspect's firearm may have been dropped at the scene. In such cases it should be left in place if this can be done safely. If safety precludes this, officers may mark the location and position of the firearm and secure it in their holster or in another acceptable manner. However, the preferred procedure is that weapons, expended cartridge casings, brass, speed loaders, magazines, and related items be left in place undisturbed.

Before back-up officers and supervisory personnel arrive, involved officers should begin to secure the area to the degree that time, resources, and individual capabilities permit. Often this is no more than securing parts of the scene that may be destroyed or damaged during the first few moments of the incident, such as evidence blowing away or being washed away by rain. If possible, the perimeter should be secured with crime scene tape or by other appropriate means, and all nonessential personnel should be precluded from entering the area. Any items of potential evidentiary value therein should be protected and back-up officers should be used wherever possible in this capacity. In some cases, emergency fire or medical personnel will need to move persons or items in order to provide medical assistance. Where this is the case, officers should note their original position and condition and provide this information to investigative officers.

Where possible, officers should also identify potential witnesses to the shooting. These individuals should be separated so that their personal perceptions can be obtained without the potential influence of opinions and observations gleaned from others. The name, address, and phone number of witnesses and other persons in the general vicinity of the shooting should be recorded. In some cases these persons will claim that they did not see anything in order not to become involved. Nevertheless, officers should attempt to collect identifying information from them so that it will be possible to contact them at a later date. Any witnesses or potential witnesses who have been identified should be asked to remain on hand until a statement has been taken from them.

Beyond performing these basic responses where possible, officers involved in a hostile-shooting incident where injury or death has occurred should prepare themselves for an extended period of sitting, waiting, and interviewing with agency investigators. Officers should not be insulted by tough questions asked by investigators following such incidents. Only by asking the tough questions can all of the facts and circumstances surrounding the shooting event be compiled.

B. Supervisory Responsibilities at the Scene

The first supervisor to arrive at the scene of an officer-involved shooting should be designated as the officer-in-charge (OIC) until such time as he/she is relieved from this responsibility by an investigator or other appropriate senior officer. The supervisor's first responsibility is to ensure that the safety and security of officers has been adequately addressed. The potential threat from assailants should be eliminated first and any suspects at the scene should be detained or arrested. Following this, emergency medical providers should be summoned if necessary and emergency first aid provided if needed in the interim.

The supervisor should ensure that the crime scene has been protected and, to the degree possible, that it is kept intact and undisturbed until criminal investigators arrive. Supervisory officers should then deal with those issues discussed in the foregoing section of this paper if officers at the scene were not able to do so. That is, supervisors may need to broadcast lookouts for suspects, request backup and related support services, secure the scene and protect any items of evidentiary value, identify persons who may have been at or within close proximity to the scene of the incident, as well as identify witnesses and request their cooperation. It is preferable to transport eyewitnesses to the station where they can be interviewed by investigators. Normally, detailed interviews with witnesses should not be conducted by supervisory personnel at the scene. If witnesses are unwilling or unable to go to the station to make a statement, the general scope of their knowledge of the incident should be established and recorded together with a record of their identification for future contact by investigators.

Supervisory personnel should be aware of the possibility that officers involved in the shooting may be suffering from post-traumatic shock as previously noted. If this is the case, they should be handled in a manner consistent with agency policy and professional practice.[2] For example, the officer(s) should be moved away from the immediate shooting scene and placed in the company of a fellow officer, preferably a peer counselor where these officers are available through the police agency.

If an officer has been shot, the OIC should ensure that another officer accompanies the injured officer to the hospital and remains with the officer until relieved. The accompanying officer should be responsible for ensuring that the clothing and other personal effects of the injured officer are not discarded but are preserved and turned over to the police department as evidence. The supervisor should ensure that the officer's family or next-of-kin is notified on a priority basis and in-person wherever possible. An in-person notification should always be made when a death has occurred. An officer should be assigned to transport immediate family members to the location where they are needed. Particular care should be taken to keep the name of the involved officer(s) from the media or other sources until the immediate family members of the officer have been notified.[3] An officer should be assigned to the family of a wounded officer in order to provide them with security, emotional support, assistance in dealing with the press, and related matters.

In addition to the notification of back-up and specialized assistance previously mentioned, supervisory personnel should contact other necessary personnel in their agency at this stage depending upon the seriousness of the incident and the requirements of their agency policy. Such notifications may include the agency's internal investigative authority, homicide investigators, chief of police or sheriff, public information officer, patrol commander, legal advisor, coroner, or chaplain.

Depending on the seriousness of the incident, it may also be necessary and prudent to establish a command post in order to better coordinate the many persons involved in the investigation. In this regard, it is also a good idea to appoint one officer as a "recorder" for the incident. The duties of a recorder are to document the event and establish a chronological record of the activities at the scene. This record should include but need not be limited to: the identities of all persons present and those who entered the incident/crime scene including emergency medical and fire personnel, actions taken by police personnel, evidence processed, and any other matters of significance.

It may also be necessary at this point to establish a media staging area. Police officer-involved shooting incidents invariably draw sizeable contingents of media representatives. If the agency has a public information officer (PIO), this individual may be used to control media representatives and provide them with information as available and appropriate. Should a PIO not be employed or readily available, the OIC will need to appoint an officer at the scene to control these individuals and to provide them with the basic details of the incident as they become available and as they are appropriate for release. Caution should be exercised in the release of any information at the scene prior to a full investigation of the incident.[4]

The supervisor should also begin collecting certain types of evidence. In some agencies, this is the responsibility of investigative officers and/or evidence technicians and, in such cases, it may be necessary to defer collection for their arrival. Irrespective of the organizational responsibilities involved, the supervisor should ensure that an appropriate crime scene perimeter has been established in order to protect evidence until collected. The pressures of time and unusual conditions at the scene of the incident may require alternative approaches to evidence collection. Bad weather, a physical location of the incident that threatens destruction or theft of evidence, the ability of officers to secure and contain the incident scene, lack of ready access to specialized personnel and related factors may necessitate immediate action to avoid the loss of evidence. Therefore, supervisory officers should be prepared to identify and gather essential components of evidence if required rather than risk their destruction or loss pending arrival of investigators. The same rationale holds true with the collection of information from bystanders, witnesses, and suspects.

Therefore, as time permits, and within the parameters of the police agency's policy and procedures, supervisory personnel should begin to document the scene or ensure that this activity is undertaken by authorized agency personnel. The overall scene should be diagrammed manually, indicating the location and relative distances between key points and items of evidence. Then, photographs and, where possible, a videotape recording should be made of the overall scene and all key pieces of evidence. Videotaping is recommended wherever possible as it provides an added perspective and dimension that still photography can't always provide. The officers' and suspects' firearms and other weapons as well as expended cartridge casings should be located, safeguarded and photographed in place. They should normally be left in place for collection by evidence technicians unless circumstances or conditions at the scene threaten to contaminate or destroy them.

At the same time, it is prudent to inspect the primary and back-up firearms of any ancillary officers who may have been at the scene during the incident to determine if their firearms have been fired recently. In so doing, one may be able to respond more accurately to potential allegations or concerns about additional shots fired by police personnel. One should also establish and mark the original position and any subsequent locations from which the officers and suspects fired.

If possible, it is also a good practice to videotape and/or photograph, in a general manner, any bystanders or onlookers who may be at the scene. Some of these individuals may be witnesses to the incident but may disperse before their identity can be determined. Some may fail to come forward as witnesses, fail to provide information if questioned or provide false identification. A visual image may assist investigators at a later time in identifying and locating these individuals, if necessary.

If the officer involved in the shooting is still at the scene of the incident and it is reasonably possible to do so, it is highly advisable to take color photographs of the officer's condition and any wounds or injuries that he/she has sustained. These photographs can provide graphic documentation regarding the extent of injuries and the actual physical condition of the officer at the scene and can provide compelling testimony to the nature and impact of the incident. Should this not be possible at the scene, such photographs should be taken with the permission of hospital emergency service providers.

C. Investigator's Responsibilities

Law enforcement agencies vary with regard to the unit assigned responsibility for the investigation of an officer-involved shooting. In some cases, this responsibility is assigned to internal affairs, officers assigned to person-to-person crime investigations in the detective division or homicide investigators. Depending on the seriousness of the shooting, the circumstances involved and the protocols of the specific agency, it may be appropriate to conduct parallel investigations of such shootings by both homicide investigators or criminal investigators and internal affairs officers. Frequently, agencies conduct concurrent criminal and administrative investigations of tactical shootings (e.g., where an officer and/or suspect is wounded or killed), the former to establish conformity with departmental policies and training and the latter to establish whether criminal conduct was involved.

Larger agencies utilize specially trained "shooting teams" to respond to such events. And, as previously mentioned, many agencies turn these investigations over to larger state or county police or sheriff's departments. The prudence of these arrangements is largely based on the experience and capabilities of the officers who staff these positions. However, as a general rule, trained and experienced homicide investigators are the best persons to conduct investigations of officer-involved shootings. Their experience generally allows them to more readily identify, organize, and evaluate relevant details of a shooting situation and to establish the facts of the event such as is the case in homicide investigations.

Investigative officers assigned the lead role in an officer-involved shooting should assume control of the shooting scene and the investigation upon their arrival. Supervisory officers and other police personnel at the scene should, from that point on, answer to the investigative OIC. The investigator should determine the degree to which the foregoing tasks discussed in this document have been completed and, where deficiencies exist, ensure that these tasks are fully completed. In particular, the lead investigator should ensure that the overall scene has been properly secured, that all evidentiary items are diagrammed, photographed, videotaped, properly recorded, collected, and stored, and that all persons present at the scene are also videotaped and/or photographed.

The next order of business is to receive a general briefing and walk-through of the scene by the initial supervisory officer or the most knowledgeable officer at the scene. A decision to include officers involved in the shooting will depend upon the knowledge of the supervisor of all details of the shooting, and the physical and emotional state of the officer(s). Investigators should ensure that all essential details of the shooting have been or are being recorded. These include the nature of the call to which the officer was responding; the time it was received and dispatched; circumstances in which suspects were encountered; the time of day of the incident, weather and lighting conditions; the names and ranks of all officers involved together with their serial numbers and assignments; the identities of all persons who have had access to the shooting scene including emergency medical service personnel and firefighters; the time of dispatch and arrival of any back-up officers; whether the officers were in uniform or, if in plainclothes, whether they were identifiable as police officers at the time of the shooting; the types of vehicles involved by officers and suspects, if appropriate; and the identities and background of all suspects and others involved in the shooting.

Throughout the investigation, the OIC should play "devil's advocate" in weighing the value of the variety of information gathered in the investigation. This includes, in particular, the statements of suspects. For example, typical suspect claims include the following:

> "I shot the officer to defend myself."
> "I grabbed the officer's gun because he beat me."
> "It was an accident."
> "I didn't know he was a cop."
> "I don't remember anything."

These common claims and excuses are starting points for critical assessment of physical evidence and the statements of others.

The same holds true with witnesses and alleged witnesses to officer-involved shooting incidents. The problems associated with the reliability of eyewitnesses and the accuracy and validity of their accounts are legendary. Steps taken to separate such individuals at the scene of shootings in order to prevent the sharing of their accounts and feelings is one step among others that can and should be taken to help maintain the integrity of these accounts. But there are no guarantees that their accounts will be trustworthy. How, for example, do you explain a deadly force incident where one or more seemingly disinterested and independent witnesses describe the shooting of a suspect who was purportedly handcuffed on the ground, only to find that physical evidence precludes this from happening? Lying is one explanation but it is not the only explanation. It is possible and has been proven that individuals' perception can be colored and influenced by their background, experiences, and predispositions. While law enforcement officers are trained to be exacting in both observation and descriptions, they are also not immune to these same problems of perception. In addition, officer judgments and perceptions can be influenced by heightened levels of fear or anxiety when operating in dangerous environments not to mention the psychological trauma of the actual shooting as heretofore described. The vast majority of officer-involved shooting investigations reveal that actions taken by officers were warranted under the circumstances. But, for example, claims by an officer that he/she believed the suspect was armed, was in the process of drawing a firearm, was holding a firearm or was otherwise posing a threat of death or serious bodily harm cannot always be taken at face value.

Careful collection and examination of physical evidence in conjunction with witness statements will generally prove sufficient to support or refute these claims and thereby focus the investigation. Some agencies conduct a criminal investigation of all tactical officer-involved shootings while others may conduct internal administrative investigations or even parallel administrative and criminal investigations. Whichever the case, the guidelines for Miranda warnings exist as in any other situation. But, questioning of involved officers should not normally include the issuance of Miranda warnings unless officers are criminal suspects. While an officer can be compelled administratively to respond to questioning or face departmental disciplinary action, any compelled testimony will be precluded from subsequent criminal proceedings. The complexity of these matters necessitates that where criminal charges appear forthcoming, the local prosecutor's office and the agency chief executive become involved before in-depth questioning of a criminal focus begins.[5]

With these issues in mind at the scene of an officer-involved shooting, investigative officers should ensure that all pertinent evidence has been or is in the process of being collected, in particular, the officer's firearm(s) and ammunition. The officer's firearm and any other firearm discharged during the incident should be taken into custody and handled as evidence.[6] If the firearm is in the possession of the officer, it should be taken in a discrete manner and arrangements should be made to replace it with another firearm or return it to the officer as soon as possible. At an appropriate juncture, the serial number, make, model, and caliber of all officer and suspect weapons used at the scene should be recorded. Expended bullets and cartridge casings should be marked, photographed in place, and eventually collected as evidence for forensic examination.

As previously noted, officer and suspect clothing can provide important (often the most important) information about the shooting and should be preserved. Often the suspect's clothes can prove the proximity of the officer(s) to the suspect; the position of the suspect's arms (either up or down); the distance and trajectory of shots that were fired; or entrance and exit points. Examination of the clothing can help prove or refute charges that the officer fired shots while the suspect was handcuffed, lying on the ground, standing with hands and arms raised, or when the suspect was not a threat to the officer. Therefore, investigators should ensure that arrangements have been made to secure this clothing before it is discarded by emergency service workers, hospital personnel, or others.

Voice transmissions are also potentially important pieces of information or evidence. Therefore, arrangements should be made during the course of the investigation to identify and interview the complaint taker and dispatcher who handled the incident and to secure and review all recorded voice and data transmissions surrounding the incident, to include the logs of mobile data terminals (MDTs) where employed.

A complete description and diagram of the shooting scene should identify, to the degree possible, the location and movement of all officers involved as well as those of suspects and witnesses, and the paths of bullets fired.

Investigative officers should also concentrate on identifying possible witnesses and obtaining preliminary statements from those persons. All such statements from these individuals should be tape recorded, as should any statements made by officers and suspects involved in the shooting. The identity of persons within the general area should be obtained as should those of persons who claim not to have seen anything. Often such persons are reluctant to speak directly to police officers at the scene while later contact may prove beneficial.

In obtaining statements, investigators should not overlook information and observations made by emergency service personnel such as paramedics and firefighters. Often, these are the first responders to the scene and have dealt directly with suspects and officers immediately following the shooting. Their initial impressions concerning the circumstances of the incident upon their arrival, what may have been said by those involved, actions taken at the scene and other matters can be of great value to the investigation.

Tape-recorded interviews should be conducted with each officer irrespective of statements made previously to supervisory personnel or fellow officers. Exceptions to this include situations where the officer has been hospitalized and is unable to respond to immediate questioning or is suffering from the effects of traumatic stress associated with the incident that is sufficient to interfere with accurate recall of events that took place. In the latter regard, investigators should be cognizant of symptoms such as time and space distortion, confusion, hearing and visual distortion, and emotional impairment as mentioned previously in this paper. Interviews with involved officers and others at the scene should be conducted as soon as possible following the incident unless such symptoms exist. These interviews should be conducted in a private location away from sight and hearing of other agency members and persons who do not have a need and a right to the information solicited from the officers. It is generally wise to remove the officer from the scene of the incident to a neutral location.[7] The officers should be advised not to discuss the incident with anyone except a personal or agency attorney, union representative, or departmental investigator until the conclusion of the preliminary investigation.

Once business at the shooting scene has been concluded, investigators may follow up on leads and additional points of contact. For example, all pertinent suspect information should be obtained, such as a complete description of the suspect, his prior criminal record to include any parole or probation history. Search warrants for suspect residences and any vehicles involved in the incident should be obtained and searches conducted where appropriate in a timely manner.

If officers have died in the shooting, investigators should ensure that appropriate steps have been taken by the agency in conformance with line-of-duty death policies and death notification procedures.[8] In such instances and where suspects have died it is particularly important to work closely with the coroner's office, to include attending the autopsies of officers and suspects. Among issues of importance are those related to the determination of entrance and exit wounds,[9] estimates of shooters' positions, the presence of any controlled substances in the decedents' blood, and related matters.

The lead investigator should brief the agency chief executive once preliminary results of the investigation have been established.[10] Normally, this should take place the day following the incident or as soon as possible. Following this, with approval of the agency chief executive or his/her designee, the investigative officer should prepare a staff memorandum that details the facts of the incident. This memorandum should be posted or distributed to all personnel as soon as possible following the shooting, preferably on the day following the incident. By doing this, staff rumors can be kept to a minimum and concerns over unknown circumstances of the event can be resolved before speculation supplants fact.

An additional briefing of the prosecutor's office is also necessary in a timely manner following the shooting incident. In some cases, a member of the prosecutor's office will respond to an officer-involved shooting as a matter of established protocol. Nonetheless, the police agency should make a preliminary statement of facts as soon as possible to the prosecutor's office and work with them closely throughout the investigation. A preliminary statement of facts may then be developed with the approval of the agency PIO and chief executive or other designated personnel for release to the press.

D. Force Review Committee

Once an investigation of the shooting incident is completed, some agencies bring these findings to a Force Review Committee, Shooting Review Committee, or similar entity within the organization that sits on an ad hoc basis to review these findings. These inquiries should not be punitive in nature as matters of criminal or civil liability or administrative punishment for involved officers should be dealt with through other established agency procedures. Rather, the scope and intent of these forums is to assess such incidents to determine whether they have any implications for the department's training function, policies, and procedures. These forums are an effort to bring together all elements of an investigation in a risk management context to improve the agency's response to these critical incidents and to make any corrections in agency practice or procedure that will help avoid identified problems in the future.

These reviews should be conducted by command level officers, personnel at the supervisory level who were involved in the incident and investigation, and any other agency specialists who can provide insight to the incident. The committee should review the reports and interview materials of involved officers, first responders, supervisors, and investigators. On initial review of this material, additional clarification may be required by Internal Affairs personnel, tactical or specialist teams, or others. Upon completion of this review, findings and recommendations may be made concerning modifications in established agency policy, training, supervision, equipment, or related matters.

Notes

1. Any officer who has been shot is at high risk of going into shock, irrespective of the severity of the wound. Self-administered first aid for persons in shock is limited and sometimes impossible. But, in some cases, there are simple steps that officers can take to minimize their chances of going into shock and maximize their chances of surviving such a critical incident. See, for example, "Emergency Care: Trauma," Training Key #375, International Association of Chiefs of Police, Alexandria, VA.

2. A full discussion of the signs and symptoms of post-traumatic shock is beyond the scope of this paper. For a complete discussion of this topic, see the Model Policy and Concepts and Issues Paper on Post-Shooting Incident Procedures, International Association of Chiefs of Police, Alexandria, VA.

3. If the officer(s) has died, the agency should adhere to policy and procedures established for line-of-duty deaths. Agencies that do not have policies and procedures in effect for these contingencies are urged to establish them as soon as possible. Agencies that wish to develop, review, and/or update such polices should contact the IACP National Law Enforcement Policy Center for a copy of the Model Policy on Line-of-Duty Death and its accompanying Concepts and Issues Paper. In addition, agencies may wish to refer to IACP Training Key #358, "Death Notification," for information and guidance on dealing with the task of informing survivors of the death of an officer or other person.

4. For a detailed discussion of information that may and may not be released at the scene of these and other incidents, see the Model Policy and Concepts and Issues Paper on Police-Media Relations, IACP National Law Enforcement Policy Center, Alexandria, VA.

5. For a detailed discussion of information concerning the rights and obligations of officers and investigators in internal administrative and criminal investigations, see: Model Policy on Complaint Review, IACP National Law Enforcement Policy Center, International Association of Chiefs of Police, Alexandria, VA.

6. The need to conduct forensic comparison of ballistic evidence in officer-involved shootings underscores the importance of obtaining ballistic samples from all officers' primary and back-up firearms as part of the agency's firearms authorization requirements. For example, officers should be authorized to carry only departmentally approved firearms and ammunition with specified loads. Ballistic samples of these firearms should be required as part of the authorization process.

7. While investigators may wish to request that the officer(s) "walk through" the incident, investigators should be aware of the impact that the shooting may have on clear recall of events by officers immediately following the incident. If a walk-through is deemed necessary, it should not be videotaped. Should the officer be suffering from faulty recall or other emotional trauma, such recordings may prove unfair and prejudicial should they be subpoenaed and entered into evidence by the plaintiff in a later civil action.

8. Op. Cite., footnote 3.

9. The importance of employing experienced officers who are completely familiar with shooting investigations cannot be overemphasized. For example, recent research has demonstrated that normal delays in reaction time of officers in hostile shooting encounters often explain unusual bullet entrance and exit wounds of suspects and others, to include seemingly unexplained and suspicious shots to the suspect/victim's back. The entrance wounds of bullets do not always provide a full explanation of the circumstances surrounding the shooting event. Only highly trained and experienced investigators can decipher the full range of potential evidence involved in a complex shooting investigation.

10. For a discussion of procedures to be used in the event that criminal or administrative charges are brought against an officer in the course of a shooting or other investigation, see for example, Model Policy and Concepts and Issues Paper on Complaint Review, National Law Enforcement Policy Center, International Association of Chiefs of Police, Alexandria, VA.

Emergency Vehicle Operations

4

Vehicular pursuits and emergency responses are highly dangerous police activities that too frequently seriously injure or kill innocent bystanders who just happen to be in the path of the fleeing suspect or a speeding police car and essentially serve as involuntary roadblocks.

It is estimated that 40 cents of every dollar paid out by governments in jury awards, settlements, and legal fees are related to the police activities of use of force, motor vehicle operations, and arrest activities.[1] A review of law enforcement officer deaths in the line of duty for the years 1998 through 2007 shows that an average of 74 officers died each year feloniously, while an average of 93 died each year in accidents.[2] A closer review of officer line-of-duty deaths from 1998 to 2007 shows that 39% of the officers were shot to death, and 45% died as a result of a motorcycle or auto accident or from being struck by a vehicle.[3]

Little attention was given to police vehicle pursuits until the middle to late 1980s when several high-profile accidental deaths occurred as a result of police pursuits over matters that the public believed to be rather trivial.

As one example, on June 24, 1984, shortly after 1:00 A.M., John Deady drove through a red light in Pasadena, Florida. A deputy witnessed the offense and gave chase. As the pursuit continued, other deputies, Pinellas Park police officers, and Kenneth City, Florida, police officers joined in the pursuit at speeds varying from 80 to 100 miles per hour. After pursuing Deady for approximately 25 miles through a densely populated urban area on highways generally frequented by heavy vehicular traffic through more than 34 traffic control devices in disregard of those signals, the pursuit reached the intersection of U.S. 19 and State Road 584, where Deady's vehicle collided with a car that was proceeding through an intersection with a green light. The broadside collision of Deady's vehicle into the vehicle crossing the intersection killed two young girls.[4] The public outrage over the deaths of two innocent young people because law enforcement officers were pursuing a person for running a red light resulted in numerous law enforcement agencies restricting their officers' discretion to engage in vehicular pursuits. Even to the layperson, the value of apprehending a traffic law violator did not justify the risk of harm the officers created by the pursuit which resulted in the tragic deaths of two innocent people.

Other similar incidents across the United States apparently influenced the AAA Foundation for Public Safety in 1995 to issue a statement about police pursuits. That statement succinctly put forth the reason police officer discretion needed to be restricted by providing the following summary of the history of police pursuits:

> In years past, the police and the public generally maintained an attitude that it was the law enforcement officer's responsibility to catch anyone evading the police regardless of the cost. It was not uncommon that officers would feel obligated to catch the violator by taking unnecessarily high risk, so that their fellow workers would not see them as derelict in their duty, or weak or cowardly. These attitudes can still influence police judgment, but the tone has been changing to a more responsible philosophy. Public, judicial, and police attitudes are changing regarding the degree of risk that we as a society are willing to take in the apprehension of a suspect. Public safety, the safety of the officer, and the safety of the violator are becoming more and more an important and persuasive element of the pursuit equation. We are beginning to realize that when a police officer engages in a pursuit, his vehicle becomes a potentially dangerous weapon, perhaps the most dangerous weapon in the police officer's arsenal.[5]

The following year (1996), the International Association of Chiefs of Police (IACP), at its 103rd annual conference, encouraged all law enforcement agencies to adopt written policies governing pursuits. They also provided a "sample" policy that all members of the IACP could use as a model in drafting their own agency policy.[6] The Model Policy developed by the IACP is attached at the end of this chapter as an appendix. The "sample" policy provided a procedure to be followed before initiating a pursuit (the pursuing officer's conclusion that the immediate danger to the officer and the public created by the pursuit would be less than the immediate danger to the public should the suspect be allowed to escape). It also recommended that the officer take into consideration the road, weather, and other environmental conditions; the population density and vehicular and pedestrian traffic; the seriousness of the offense; and other factors before initiating a pursuit.[7] Unfortunately, the effectiveness of a pursuit policy that limits the discretion of an officer to engage in and continue a pursuit is not easy to evaluate, because the number of pursuits engaged in by the police annually is not measured, and the number of pursuits that did not occur because a policy has been put into place that limits an officer's discretion to engage in a pursuit is not measureable.

Notwithstanding difficulties in measuring effectiveness, most police agencies now have adopted restrictive pursuit policies that attempt to balance the risks of a pursuit against the need to immediately apprehend a suspect who is too dangerous to remain at large.

Opponents of restrictive policies argue that pursuit bans encourage lawlessness, and supporters of restrictive policies argue that adopting such policies reduces citizen and officer injuries and deaths and reduces the risk of lawsuits. In either case, the effects of the policy decision are large. When Metro-Dade (Miami), Florida, adopted a "violent felony only" pursuit policy in 1992, the number of pursuits decreased by 82% the following year. In 1993, after Omaha, Nebraska, changed to a more permissive policy, permitting pursuits for offenses that had previously been prohibited, the number of pursuits increased by 600% the following year.[8]

As with most policies, the official statement of how business is to be conducted must be supplemented by training, supervision, and accountability. In the aftermath of a pursuit that ends in tragedy, it is often found that training in the policy was minimal, supervision was lacking, and excusing a pursuing officer's continuation of a dangerous pursuit for a nonviolent crime was substituted for accountability. Uniquely in the State of California, pursuing officers and their departments are immune from liability to innocent third parties who are injured or killed by a fleeing driver, if they adopt a state-prescribed pursuit policy. The Catch 22, however, is that case law has since held that such a policy need only be *adopted* and need not be *followed* (see *Kishida v. State of California*). There can be little doubt that a comprehensive pursuit policy that is followed will prevent accidents and save lives, but when a pursuing officer is pumping adrenaline and becomes "caught up in the spirit of the chase," policy limitations tend to be stretched or ignored.

What Would You Have Done?

The scene is a major metropolitan area of 5 million population served by a central city police department, several smaller suburban police departments, and a sheriff's office for the surrounding unincorporated areas. There are three separate dispatch frequencies that are unable to directly communicate with one another. Although the basic emergency vehicle operation course (EVOC) training provided by the state is similar, each agency has a local policy manual. Accordingly, each agency has an individual pursuit policy but all require that any pursuit must balance the inherent risks against the need to immediately apprehend a fleeing suspect.

Jason Adams was 19 years of age and had a driving record that resulted in his driver's license having been suspended, and he had two outstanding misdemeanor warrants for having failed to appear on his last two traffic citations. He felt that he could not afford to be stopped by the police again, and he was driving carefully so as to not attract any unwanted attention. Unfortunately, he did not know that his brake lights were out, when he stopped for a traffic signal, and a deputy sheriff, who was on routine patrol that Saturday evening at 10:30 P.M., noted the equipment violation. The deputy decided to stop the driver for a warning, and after positioning his patrol vehicle directly behind him, he waited for the traffic signal to change. When it soon did, the deputy activated his overhead lights to make the stop and saw the driver immediately glance at his rearview mirror but not slow down. Jason was not sure what to do next, but he knew that he did not want to be arrested or go to jail and continued to drive, while his mind raced through his options. After a short distance farther, the deputy activated his siren, whereupon Jason panicked, immediately accelerated away, and the chase was on. The deputy reported to his dispatcher that he was in pursuit of a traffic violator, who would not stop, gave his location and direction of travel, and started "calling the pursuit." Speeds soon reached in excess of 80 mph, on the two-lane road where the pursuit started, and oncoming vehicles were pulling over onto the shoulder to move out of the way. When Jason caught up to any traffic ahead of him, he either passed on the right or crossed into the oncoming lane and forced traffic off of the road. There was a red traffic signal, where the pursuit reached a major arterial, and Jason ran the red traffic signal, slowing only enough to skid through the turn, while other vehicles slammed on their brakes to avoid an accident. By now, adrenaline had kicked in to affect the perceptions and attitudes of both the pursued and his pursuer. Jason's panic had him focused on his rearview mirror to see if he was still being chased, and the pursuing deputy's focus had narrowed to the fleeing vehicle, as "tunnel vision" started to take effect. The dispatcher had already noticed that the deputy's voice was sounding increasingly stressed, as he called out speeds, locations, and the fleeing vehicles near accidents with other motorists. At one point, the dispatch audiotape recorded him saying, "This guy is going to kill somebody." As the pursuit continued down the arterial toward the central city, Jason passed through one of the suburbs, and an officer from that city's police department joined the pursuit as the secondary unit. Because of the different dispatch frequencies between the sheriff's office and the suburban police department, the suburban officer could not communicate with the deputy and, in fact, did not even know the reason for the pursuit. Speeds continued upward in excess of 100 mph, and as more controlled intersections were encountered, Jason ran every red signal without slowing in his desperate attempt to escape at any cost. The pursuing deputy remembered wondering how many of those intersections the fleeing driver could make it through before he hit someone. In fact, his report made a comparison to "Russian roulette" and questioned, "How many times will he try again (pull the figurative trigger) before he hits the loaded intersection (figurative chamber)?" Nevertheless, the pursuit continued unabated with no thought to termination or any supervisory intervention, and the deputy becoming ever more determined to catch the fleeing driver, no matter what. After the pursuit had traveled 9.7 miles and accumulated a few more pursuing police units, it approached the downtown area of the central city. Unaware of the approaching pursuit and neither seeing emergency lights nor hearing a siren, a young family was driving home, with their windows rolled up and the radio playing. Not surprisingly, and as he had done several times before, Jason ran yet another red traffic signal, but this was the "loaded" one occupied by the young family's vehicle that was crossing the pursuit's path with the right of way, when they were T-boned at 90 mph. The innocent occupants of that vehicle were a father, who was killed instantly, a mother who died en route to the hospital, a 10-year-old girl, who was rendered a quadriplegic, and an infant boy, who was uninjured. Jason survived the accident to be charged with vehicular assault and two counts of vehicular homicide. And, of course, he was cited for the defective brake lights that had started the whole fatal event. The media reported the tragic accident, and the sheriff's office soon released a press release stating, "The pursuing deputy complied with their pursuit policy because the fleeing driver was so dangerous that he had to be stopped."

Investigation of Accidents

There is a tendency to treat a collision at the end of a police vehicular pursuit as just another accident, but it is not. Accidents resulting from such pursuits are the source of significant liability for the pursuing agency and should be investigated with the same objective thoroughness as an officer-involved shooting. It is essential that the accident investigation be conducted by a certified accident reconstructionist, and if the agency of jurisdiction does not have such a capability, the state police or highway patrol should be asked to provide that necessary assistance. The accident scene must be protected until the reconstructionist can complete the preliminary gathering of evidence by use of a total station or similar precision measuring device to accurately locate and document all relevant facts. In addition to the thorough and detailed documentation of a professional accident reconstruction, which need not be repeated here, it is additionally important to preserve the audio recording of all radio transmissions preceding, during, and immediately after the pursuit. Failing to specifically preserve such evidence will allow the recording to be recycled and critical evidence of how the pursuit was actually conducted will be irretrievably lost. Unlike most accident investigations, an accident at the end of a vehicular pursuit begins with the precipitating event and continues through the chase until it is terminated in a collision. Accordingly, the investigation will include an analysis of how the pursuit was actually conducted and whether or not it was in compliance with policy and training. Remember that the purpose of a vehicular pursuit accident investigation and review is threefold:

1. Document how the accident occurred.
2. Determine if policy and training were followed.
3. Prevent future officer-involved deaths from vehicular pursuits.

Policy and training in vehicular pursuit are designed to prevent fatal or injurious accidents that cannot be justified under the totality of circumstances. A thorough investigation and review of those that were not prevented can only serve to make the future safer for all concerned.

The scenario in this chapter is all too common. Vehicular pursuit is one of the most dangerous police activities and causes more injury and death than any other police function. A defective brake light led to a hot pursuit and ended up with the injury of a 10-year-old girl and the death of her parents—an innocent young couple. The driver, Jason Adams, more likely than not made a deal with the district attorney and plead for a lesser charge than homicide. The pursuing deputy and other police officer and the involved police department all ended up with a multimillion dollar civil law suit. All of this started with a defective brake light.

Figure 4.1 depicts an aerial view of a pursuit scene. This high-speed police pursuit ended with the suspect vehicle being boxed in by police vehicles. The driver was killed by the pursuing police officer.

In these types of situations, the shift supervisor should take over the pursuit action and make a decision whether to call off the pursuit according to the department policy. Of course, if unfortunately the tragedy has already happened, ambulance and additional backup personnel should be dispatched right away. The shift supervisor or commanding officer should respond to the

Figure 4.1 An aerial view of the end of a high-speed pursuit scene. (**See color insert following page 78.**)

scene immediately. Fire department and emergency medical services (EMS) should also dispatch to the scene to start rescue functions. A multiagency joint investigation team should assemble to investigate the accident, ensuring that these tasks take place:

- EMS should remove and transport the injured individuals to a hospital as soon as possible.
- Fire department personnel should assist rescue and prevent any vehicular fire and explosion.
- Crime scene personnel should secure the crime scene and begin the crime scene protection measures through the use of barrier tape, official vehicles, or, if necessary, by closing the road.
- Law enforcement personnel should initiate traffic control and direct the traffic flow. If necessary, they should complete reroutes of all traffic.
- Law enforcement personnel should establish a command post in accordance with departmental operations policies and notify the appropriate agencies to initiate a multiagency investigation team.
- Crowd control activities should be undertaken by law enforcement personnel to protect the integrity of the scene and to prevent additional traffic accidents by passing motorists, causing the possible escalation of the event.
- Search for witnesses remembering the pursuit. (This covered an area over 9.7 miles.) Potential witnesses may have valuable information about the incident. Those witnesses need to be located and interviewed.
- Remove officer(s) involved in the pursuit from the scene to avoid potential confrontations between the officer and suspect family members.
- Officer statements should also be taken as soon as possible.
- Initiate Crime Scene Investigation Procedures. This scene could consist of multiple subscenes along the 9.7 miles of roadways.
- Notify the appropriate agencies, including the traffic accident squad, forensic laboratory, medical examiner, district attorney's office, and other law enforcement agencies when required.
- Notify the decedent's family.

Crime Scene Investigation

General crime scene procedures should be followed. Because this incident covered approximately 9.7 miles of roadway and went through several police jurisdictions, it is important to select one agency to be the lead investigation agency. Figure 4.2 shows state troopers and forensic investigators working at a fatal accident scene after a police pursuit of a speeding vehicle.

Figure 4.2 A Connecticut State forensic team investigate a high-speed pursuit accident scene on I-95. **(See color insert.)**

I. Crime Scene Survey

The 9.7 miles of roadway could be divided in sections according to jurisdictions. Each agency team would then cover their area of jurisdiction to conduct a scene survey and to locate potential evidence and witnesses.

II. Documentation of the Crime Scene

A. Each scene should be thoroughly documented using the standard crime scene documentation techniques. Figure 4.3 shows a multiagency task team working together at a crime to document and collect forensic evidence. Each piece of physical evidence should be assigned a number and close-up photographs should be taken. The following areas are especially important for reconstruction of the event.

B. Documentation of each scene involved in the pursuit, including:

a. Photographs and measurements of all patterns and injuries

b. Photographs and documentation of vehicle damages

c. X-ray factures and determine the impact direction

d. Identify patterns imparted from the vehicle

e. Trace evidence on victim's clothing

f. Trace evidence on vehicle

g. Transfer evidence and tire marks at the scene

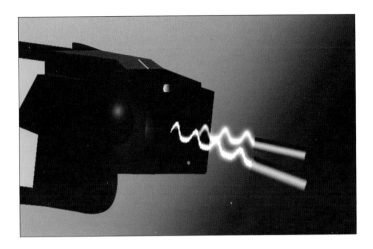

Figure 2.1 A close-up view of a conducted energy device (CED).

Figure 3.1 An overall view of a shooting death scene.

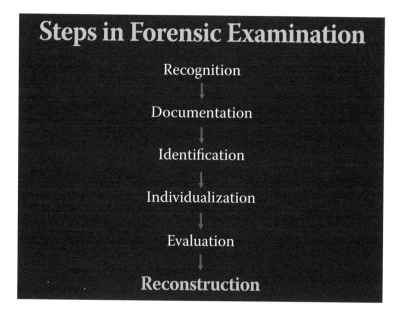

Figure 3.2 Steps in forensic examination of a crime scene.

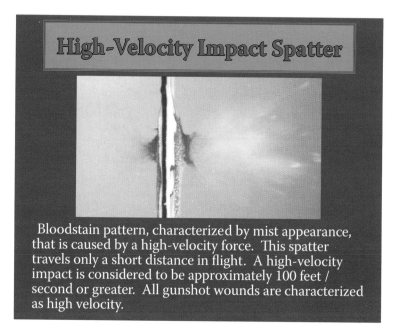

Figure 3.3 High-speed close-up photograph of high-velocity impact blood spatter.

Figure 3.4 Close-up view of the biological evidence found on the gun barrel.

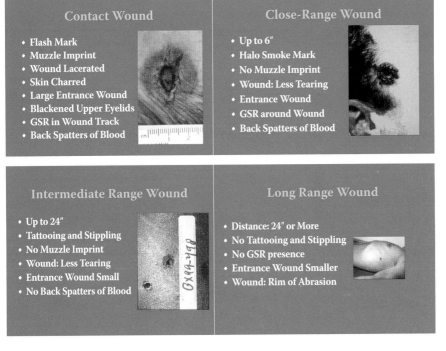

Figure 3.5 Gunshot wound pattern analysis.

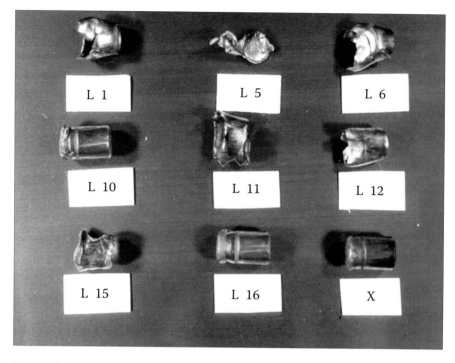

Figure 3.6 Bullet deformations and damage patterns.

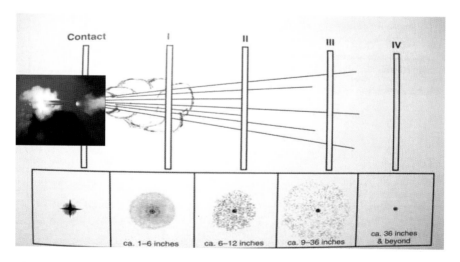

Figure 3.7 Gunshot residue pattern analyses.

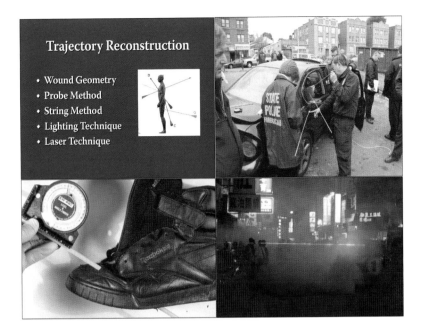

Figure 3.8 Techniques used for bullet trajectory reconstruction.

Figure 4.1 An aerial view of the end of a high-speed pursuit scene.

Figure 4.2 A Connecticut State forensic team investigate a high-speed pursuit accident scene on I-95.

Figure 4.3 Investigators reconstruct a pursuit accident scene.

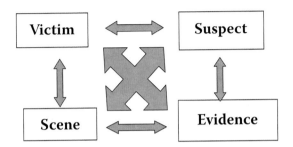

Figure 4.4 Four-way linkage theory.

Figure 4.5 Pattern evidence found on the exterior surface of a vehicle involved in an accident.

Figure 4.6 Biological evidence found in the interior surface of a vehicle involved in an accident.

Figure 4.7 Tire marks left at a fatal car accident.

Figure 4.8 Forensic team reconstructs an accident scene to determine the speed of the vehicle.

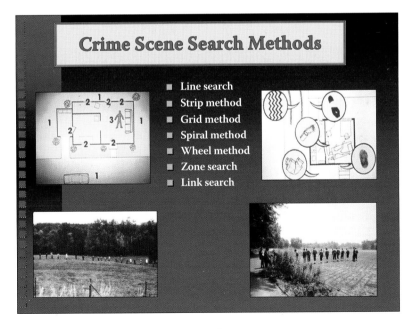

Figure 5.1 Methods of crime scene search patterns.

Figure 6.1 Crime scene investigator thoroughly documenting the scene.

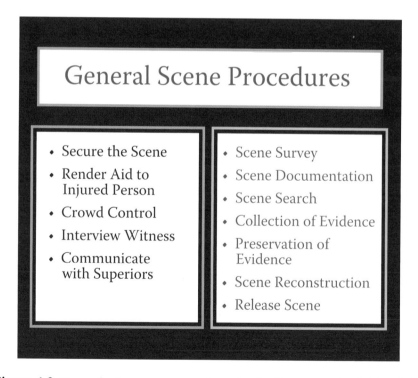

General Scene Procedures

- Secure the Scene
- Render Aid to Injured Person
- Crowd Control
- Interview Witness
- Communicate with Superiors

- Scene Survey
- Scene Documentation
- Scene Search
- Collection of Evidence
- Preservation of Evidence
- Scene Reconstruction
- Release Scene

Figure 6.3 General crime scene procedures for the first responder (left) and the crime scene investigator (right).

Figure 7.1 Forensic investigators examining an indoor scene.

Figure 7.2 Physical evidence located at a restraint asphyxia death scene.

Figure 7.3 Trace evidence found at a crime scene.

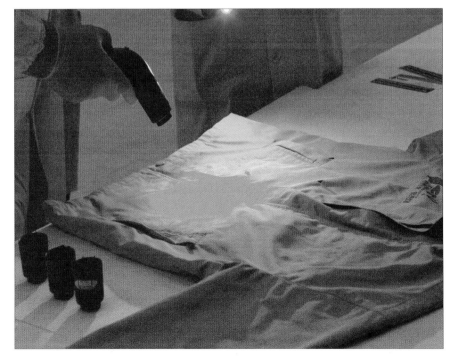

Figure 7.4 A laboratory scientist examines a shirt.

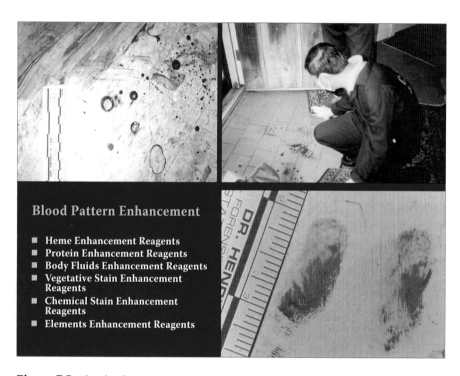

Figure 7.5 Blood enhancement reagent used to develop pattern evidence at a crime scene.

Figure 7.6 Officers engaged in a restraint asphyxia scenario.

Figure 8.1 A jail suicide.

Figure 9.1 News articles on deaths related to police activity.

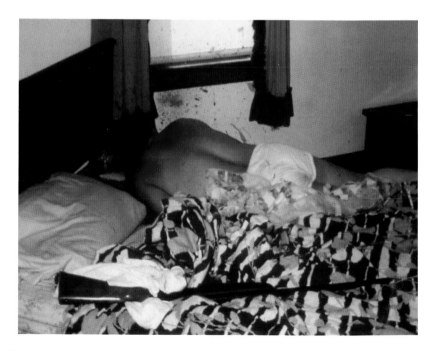

Figure 9.2 A view of an indoor shooting scene.

Figure 4.3 Investigators reconstruct a pursuit accident scene. (**See color insert.**)

III. Crime Scene Search

   A. A search should be carried out on each scene according to the size and condition of the scene. The purpose of the search is to locate potential physical evidence and to reconstruct the events leading to the fatal accident.

   B. The link method is often the most productive and common approach for this type of scene. This method is based on the four-way linkage theory, seeking to find associations between the scene, victim, suspect, and physical evidence. With this method, the investigators evaluate the scene(s) and then proceed through the areas in a systematic and logical fashion to gather physical evidence that can be linked or associated to a particular crime or activity. Linkage Theory is a concept used to explain the inter-relations between a crime scene (including vehicles), a victim, a suspect, and physical evidence. Understanding and appreciating these connections between these components will provide guidance in determining where evidence may be located, and the need to identify evidence so that the linkages may be established. Figure 4.4 shows the basic principle of this four-way linkage.

   Theoretically, if associations can be established between the damage to vehicles, debris and paint chips found on each

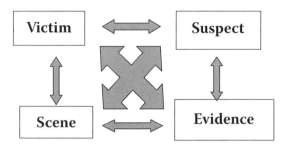

Figure 4.4 Four-way linkage theory. (**See color insert.**)

scene, tire tracks and skid marks along the road, injury patterns to victims and suspects, the incident may be reconstructed and resolved. The more associations established, the greater the probability of successfully determining the facts about the event. For example, if the suspect claims he never drove over 100 mph during the pursuit, the skid marks and the extent of damage to the vehicles could prove that the suspect was lying. At the same time, evidence such as metal fragments, paint chips, vegetative material, soil, or glass collected from the scene may establish a link between different vehicles at different scenes.

IV. Collection, Preservation, and Packaging of Physical Evidence

  A. In this particular scenario, the following types of physical evidence should be collected:

    a. Cruiser exterior and interior, damage to windshield, bloodstains, vehicle identification number (VIN)

    b. Year, make, model, color, style, type

    c. Damage classification: disabling, function, monitor, no damages

    d. Inspect and record the exterior surface of the vehicle

    e. Inspect and record the interior of the vehicle

    f. Multiple contact

    g. Light switch functional condition

    h. Speed and odometer reading

    i. Alcohol or drug containers

  Figure 4.5 shows the exterior view of a vehicle involved in a fatal chase. Figure 4.6 shows the interior of the same vehicle.

V. Evidence from Jason Adams

  A. Clothing and shoes

  B. Injury patterns

  C. Medical and autopsy reports

  D. Toxicology results

Figure 4.5 Pattern evidence found on the exterior surface of a vehicle involved in an accident. **(See color insert.)**

Figure 4.6 Biological evidence found in the interior surface of a vehicle involved in an accident. **(See color insert.)**

Figure 4.7 Tire marks left at a fatal car accident. (**See color insert.**)

VI. Evidence at Scene(s)

Tire marks: the following types of tire mark evidence may be found on road surfaces. Figure 4.7 shows tire marks left behind by the fatal chase accident scene. These marks are valuable forensic evidence for determining vehicle speed, direction of travel, action during the event, and reconstruction of the accident.

A. Tire friction marks: Made when a slipping or sliding action takes place

B. Tire imprint marks: Made without sliding on road surface

C. Skid marks: Sliding without rotation of wheel (to determine if vehicle braked before the collision)

D. Acceleration scuff marks: Made by both sliding and rotation

E. Paint chips and paint smears on vehicles

F. Damage and indentation on each vehicle

G. Bloodstains and their patterns

H. Body positions and locations

I. Windshield fracture patterns and glass fragments

J. Seat belts and air bag conditions

K. Trace and transfer evidence, such as soil, plastic, metal

VII. Preliminary Reconstruction

One of the most important aspects of the investigation is to reconstruct the event. Reconstructing the location and position of the fatal impact and determining the speed and brake time of the impact are essential

Figure 4.8 Forensic team reconstructs an accident scene to determine the speed of the vehicle. (**See color insert.**)

aspects of the investigation. This information could be used to confirm or refute the statements from witnesses, suspects, and police officers. Figure 4.8 shows forensic scientists and crime scene investigators working together at the scene to calculate the speed of the vehicle, to determine the impact point, and to reconstruct a fatal police pursuit scene.

VIII. Releasing the Scene

Because some scenes are located on heavily traveled roadways, it is essential to have a proper traffic control plan. The scene should be processed as quickly as possible. All the involved vehicles should be impounded and secured in an indoor area for further processing by a forensic team.

Best Practices

It is absolutely essential that every police agency adopt a restrictive pursuit policy and that pursuits be strictly limited to known suspects of violent felonies:

I. Pursuits of stolen vehicles are not cost effective:
   A. The auto thief will almost always run from the police and is typically a young and inexperienced driver fleeing in sheer panic at any cost.

B. Stolen vehicle chases account for approximately half of the pursuit-related accidents.

C. It serves little purpose to recover a wrecked vehicle at the risk of death or serious injuries to other drivers or pedestrians.

II. Although DUI (driving under the influence) enforcement is an important responsibility that usually begins with, "We've gotta get 'em off the road," we must also consider the following dilemma:

A. If the DUI will not stop, when signaled to do so, how are you going to make him stop?

B. Was the DUI more dangerous weaving his way home or has he become more dangerous in panicked high-speed flight from arrest with his eyes glued to the rearview mirror?

C. Not a simple question, of course, but the issue becomes one of how can we best stop admittedly dangerous behavior, without *causing* that behavior to become exponentially more dangerous?

III. The agency's pursuit policy must be followed in actual practice, with the same emphasis as is given to the agency's deadly force policy:

A. Officers and supervisors must be thoroughly trained:

  a. They must know and understand policy limitations.

  b. As with firearms training, officers must be not only trained in how to pursue, but they must also be trained in the decision-making process of when and when not to pursue.

  c. Remember, no matter how well the pursuing officer is trained in driving skills, none of that training is somehow transferred to the fleeing driver.

B. Officers must be supervised during a pursuit:

  a. Relying upon a pursuing officer to exercise *good judgment* during the heat of pursuit is unlikely to yield the desired results.

  b. A supervisor must be immediately advised of the reason for the pursuit and make an objective (i.e., uninvolved and dispassionate) decision to continue or terminate.

C. Officers and supervisors must both be held accountable for policy compliance:

  a. All pursuits, regardless of outcome, must be reported and analyzed for policy compliance.

  b. If officers and supervisors are not held accountable for policy compliance, the agency's real practice will be to pursue anyone that flees.

IV. Understand the fallacy of two myths that have prevented a reduction of the deaths and serious injuries that continue to result from pursuits that should not have occurred:

A. "If we do not pursue everyone who flees from the police, everyone will flee." The fact is that those suspects who would flee from the police are going to continue to do so, and those who would not flee still will not—why should they?

B. "Suspects who flee from the police are wanted for some heinous crime that will be discovered after they are caught." The fact is that the fleeing driver does not have his or her latest murder victim in the trunk, and nearly all serious crimes will be cleared, as they always have been, by investigation and not a happenstance encounter on a traffic stop.

Pathology

Unlike most police-involved deaths, an accident at the end of a police vehicular pursuit is the obvious cause of death, and pathological analyses will add little to reconstruction of the accident. The exception, of course, is toxicology, and it is important to know the level of alcohol or drug intoxication of the pursued driver or an incident in which the driver of the fleeing vehicle likely died as a result of something other than trauma. For that reason, an autopsy of an individual killed from gunshot while fleeing from the police is provided.

Autopsy Report of Police Shooting during Pursuit

The body is that of a well-developed, adequately nourished African American male, weighing 187 pounds, measuring 73.5 inches, and appearing to be the stated age of 44 years.

The body is dressed in a black jacket; brown, red, and white "terry cloth" shirt; blue jeans with a black and red cloth belt; white thermal underwear, top and bottom; two pairs of socks, gray and white; and brown, ankle-high boots. The jacket, shirt, and thermal top show multiple defects corresponding with the gunshot wounds of the body. Blood is noted on the jacket, shirt, and upper part of the thermal top and pants.

Note: Two bullets are found in the shirt. These bullets are placed in an appropriately labeled envelope with the name of the deceased, to be submitted to the County Crime Lab.

The temperature of the body is cold to the touch. Rigor mortis is well developed and present to an equal extent in all the joints. Livor mortis is unrecognizable due to natural skin pigmentation. The skin is pale, dry, and smooth.

The head and face exhibit trauma, which will be described below. The head hair is braided and black with a few gray hairs. The eyes are brown with pale conjunctivae. The corneas and lenses are transparent. No petechial hemorrhages or congestion are noted in either conjunctiva. The pupils are regular, round, equal, central, and measure 0.5 cm in diameter. The ears and external auditory canals are unremarkable. The skeleton of the nose is intact, and no foreign material is present in the nostrils. No foreign material is present in the oral cavity. The upper and lower teeth are natural and in a fair state of dental repair. One of the upper frontal teeth is absent. A mustache is present.

The neck is symmetrical and exhibits trauma, which will be described below.

The shoulders are symmetrical.

The chest is symmetrical and exhibits injuries, which will be described below.

The abdomen is flat. No masses can be palpated through the abdominal wall.

The back is symmetrical and exhibits trauma, which will be described below.

The external genitalia and the anus are unremarkable. The penis is circumcised, and the testes are present in the scrotum.

The extremities are symmetrical and exhibit trauma, which will be described below. The fingernails are evenly cut, dirty, and short. The toenails are short and unremarkable. No edema is present in the ankles or legs.

Evidence of Recent Trauma

    I. Gunshot Wound to the Back of the Head
       A. Gunshot wound entrance:
          A 0.4-cm-diameter skin defect is noted on the back of the head, 1 cm to the left of the midline and 3.5 cm below the top of the head. The wound is surrounded by a 0.1-cm rim of abrasion. No soot or powder stippling is noted around the wound. A sample from the hair surrounding the wound is pulled and will be submitted to the County Crime Lab.
       B. Path of the bullet:
          The bullet perforates the skin and subcutaneous tissue and then enters the cranial cavity after perforating the skull, creating an inwardly beveled hole. The bullet then passes through the meninges and the left cerebral hemisphere to settle at the left side of the middle cranial fossa.
       C. Site of recovery of the bullet fragments:
          Bullet fragments are recovered from the left side of the middle cranial fossa. Another bullet fragment is recovered from around the gunshot wound of entrance during the reflection of the scalp. The fragments are placed in an appropriately labeled envelope with the name of the deceased and will be submitted to the County Crime Lab.

D. Direction of the bullet:
The trajectory of the bullet is forward and to the left.

II. Gunshot Wound to the Right Side of the Head

A. Gunshot wound entrance:
A 0.6-cm skin defect surrounded by a 0.2-cm rim of abrasion is noted on the right parietal region of the scalp, 3.2 cm to the right of the midline and 9 cm above the right ear. No soot or powder stippling is noted around the wound.

B. Path of the bullet:
The bullet perforates the skin and subcutaneous tissue and then enters the cranial cavity, creating an inward beveling. The bullet then passed through the meninges and both sides of the brain, settling in the left occipital lobe.

C. Site of recovery of bullet fragments:
Bullet fragments are recovered from the left occipital lobe of the brain. The fragments are placed in an appropriately labeled envelope with the name of the deceased and submitted to the County Crime Lab.

D. Direction of the bullet:
The trajectory of the bullet is right to left and backward.

III. Gunshot Wound to the Neck

A. Gunshot wound entrance:
A 0.6-cm skin defect is noted on the right side of the back of the neck, 3 cm to the right of the midline and 18 cm from the top of the head. The wound is surrounded by an eccentric rim of abrasion with its widest side toward the right, 0.8 cm from the border of the wound. No soot or powder stippling is noted around the wound.

B. Path of the bullet:
The bullet perforates the skin and subcutaneous tissue and fractures the neural arch of the fourth cervical vertebra. The bullet then passes through the paravertebral muscles and exits the neck from the left of the back.

C. Gunshot wound of exit:
A 1.5 × 0.8 cm wound is noted on the left side of the back of the neck, 4.5 cm to the left of the midline and 20 cm from the top of the head. The wound shows no soot or powder stippling around it.

D. Direction of the bullet:
The trajectory of the bullet is right to left and slightly downward.

IV. Gunshot Wounds to the Right Shoulder and the Right-Upper Quadrant of the Back (11)

A. Gunshot wound entrances:

Eleven gunshot wounds of entrance are scattered on the lateral aspect of the right shoulder and right upper quadrant of the back, ranging in distance from the top of the head between 22 cm (gunshot wound 1) and 19 cm (gunshot wound 11), and to the right of the midline from 7.5 cm (gunshot wound 1) and 22 cm (gunshot wound 9) from the posterior midline. The wounds are each surrounded by a rim of abrasion. In the majority of the wounds, the rim of abrasion is eccentric with the widest portion toward the right as in gunshot wound 10, where it is 1.3 cm in width. Two pairs of wounds show close proximity to each other; these are gunshot wounds 4 and 5 and gunshot wounds 6 and 7. No soot or powder stippling is noted around the wounds. A foreign material (cloth) is recovered from gunshot wound 6 and submitted to the County Crime Lab.

B. Path of the bullets:

The bullets perforate the skin and subcutaneous tissue. Four of them enter the chest cavity fracturing the first, second, fourth, and fifth ribs. Two of the bullets perforate the right and left lungs with one settling in the pleural cavity and one in the fourth left intercostal space. The other bullet passes through the fourth thoracic vertebra, fracturing the fifth thoracic vertebra and lacerating the back of the left ventricle. One bullet is recovered from the left paravertebral muscles of the upper thoracic vertebrae. Four bullets pass through the muscles of the back fracturing the first and third thoracic vertebrae, settling in the neural arches of the first and third thoracic vertebrae after transecting the spinal cord. Another bullet lacerates the left paravertebral muscles and settles behind the left clavicle. A bullet jacket and a bullet are found behind the right clavicle. Three bullets exit the chest—two through the back and one through the front.

C. Gunshot wound exits (3):

Irregular wounds ranging between 1.5 × 1 cm and 1 × 0.3 cm are noted: one is on the front of the chest, 32 cm below the top of the head and to the left of the midline. The other two wounds are noted on the back 23 cm and 28 cm below the top of the head and 14 cm and 4.5 cm to the left of the midline of the back, respectively. No soot or powder stippling is noted around them. One of the wounds shows a piece of cloth on it (gunshot wound 12).

D. Site of recovery of the bullets:

Seven bullets and one jacket are recovered from the chest cavity and wall as mentioned previously. All the recovered bullets are placed in an appropriately labeled envelope with the name of the deceased and submitted to the County Crime Lab.

E. Direction of the bullets:

The majority of the bullets traveled from the right to the left. Some have a forward direction. One bullet has a forward direction and exits from the front of the chest.

Other Evidence of Trauma

1. A 0.8 × 0.4 cm red abrasion is noted on the lower lip, right side.
2. A 3.5 × 0.8 cm red abrasion (consistent with bullet grazing) is obliquely oriented on the anterolateral aspect of the right arm.
3. A 1 × 0.5 cm scabbed abrasion is noted on the medial aspect of the back of the lower third of the right arm.
4. A multiple, linear, scabbed abrasion is noted on the anteromedial aspect of the upper part of the right forearm, measuring 11.5 × 8 cm.
5. Multiple, scabbed, linear abrasions measuring 9 × 5 cm are noted on the upper third of the ventral aspect of the left forearm.
6. A 2.2 × 0.3 cm dark purple contusion is noted on the middle of the upper part of the back.
7. Two red contusions, 1 × 0.3 cm and 0.5 × 0.3 cm, are noted on the right side of the base of the neck.

Internal Examination

Body Cavities

The body is opened by a "Y"-shaped incision. The abdominal fat pad is 2 cm thick at the umbilicus. The muscles of the chest and abdominal wall are normal in color and consistency. The sternum and spine exhibit no fractures. There are fractures of the right first, second, fourth, and fifth ribs and upper border of the left fifth rib. The pleural cavities are smooth, and each cavity contains 1000 cc of blood. The peritoneal cavity is dry. The liver and spleen do not extend below the costal margins. The bladder lies below the symphysis pubis. The organs of the pleural and peritoneal cavities are in their usual positions in situ. The mesentery and omentum are unremarkable.

Neck

The gunshot wound of the back of the neck was described previously. The soft tissues of the neck, thyroid, and cricoid cartilages, larynx, and hyoid bone show no hemorrhage or evidence of traumatic injury, except for the gunshot wound to the back of the neck. The laryngeal mucosa is pink. The epiglottis and vocal cords are unremarkable.

Cardiovascular System

The gunshot wound of the chest was described previously. The heart weighs 345 grams. The pericardium contains a few cc of blood. The epicardial surface is smooth. The external configuration of the heart shows the previously mentioned nonpenetrating 4-cm laceration of the left ventricle. A 3.5 × 1.5 cm contusion is also noted near the laceration. The right and left ventricles are unremarkable. The endocardium and valve leaflets are smooth and transparent and exhibit no thrombi, vegetations, or fibrosis. The trabeculae carneae and papillary muscles are unremarkable. The chordae tendineae are usual. The right ventricle is 0.3 cm thick, and the left ventricle is 1.5 cm thick. The coronary arteries have their usual distribution. The coronary ostia are normal in patency. Multiple cross sections at 0.2-cm intervals show minimal atherosclerotic changes with no significant narrowing. The myocardium is of usual consistency, dark brown, and homogeneous.

The aorta shows a 2.5-cm irregular perforation at the middle of its thoracic portion.

The venae cavae are unremarkable.

Respiratory System

The gunshot wounds of the lungs were described previously. The right lung weighs 455 grams, and the left lung weighs 345 grams. The tracheal mucosa is unremarkable. The pleurae are delicate and glistening and show the previously described bullet perforations. The lungs are not distended and have a variegated pink-gray to dark purple color. The lung parenchyma is of usual consistency and mottled with a moderate amount of anthracotic pigment. The lung tissue is mildly congested and edematous. No nodularity is seen.

The extra and intrapulmonary bronchi are unremarkable. The pulmonary arteries and veins exhibit no pathological change. The hilar and mediastinal lymph nodes are unremarkable.

Hepatobiliary System

The liver weighs 1500 grams. The capsule of Glisson is transparent. The external surface is smooth, glistening, and brown. The borders are sharp. The parenchyma is soft and brown with the usual lobular architecture and no focal or diffuse lesions.

The gallbladder has delicate walls and contains a few cc of yellow bile and has a smooth mucosa. No stones are present.

The intra- and extrahepatic biliary ducts are patent. The hepatic and portal veins and the hepatic artery are unremarkable.

Hemolymphatic System

The spleen weighs 95 grams and is soft. The capsule is glistening and purple. The internal architecture is clearly defined.

Gastrointestinal System

The esophagus is empty and unremarkable. The stomach is filled with partially digested food. The remainder of the gastrointestinal system is unremarkable.

The appendix is not identified.

Urogenital System

The right kidney weighs 125 grams, and the left kidney weighs 125 grams. The surfaces are smooth and glistening. The capsules strip easily, revealing red-brown surfaces. The corticomedullary junctions are well defined. The calyceal and collecting systems are not remarkable. The renal arteries and veins are unremarkable.

The ureters are not dilated or obstructed.

The bladder contains clear urine. The bladder exhibits the usual mucosa and muscularis. The ureteral orifices are patent.

The prostate is not enlarged and does not constrict the urethra. The tissue of the prostate is lobulated, tan, and moderately firm.

Endocrine System

The adrenals, thyroid, parathyroids, pancreas, and pituitary are not remarkable.

Musculoskeletal System

There are no gross bony deformities. The muscles are well developed and show the usual color and consistency. The sternum, ribs, and spine exhibit usual bone density and marrow.

Central Nervous System

The gunshot wound of the head was described previously. The scalp is reflected, revealing focal subcutaneous and subgaleal hemorrhage. The calvarium shows linear fractures, and upon removal, there is evidence of minimal epidural and subdural hemorrhages. The dura mater does not exhibit any stains but shows the bullet perforations previously described. The leptomeninges are not remarkable, except for minimal hemorrhage around the bullet perforations.

The brain weighs 1220 grams and is of usual consistency. The undamaged sulci and gyri occupy their usual positions and exhibit a normal depth. The blood vessels at the base do not reveal any aneurysms. The cerebral and cerebellar hemispheres are symmetrical, and the surface does not display any scar tissue. The ventricles contain the usual amount of colorless fluid. The cerebellar tonsils are not herniated. Multiple sections through the cerebrum, cerebellum, pons, midbrain, and medulla exhibit the usual internal pattern with the previously mentioned lacerations.

The skull shows a fracture of the vault and base caused by the bullet perforations.

Note: Blood, bile, urine, eye fluid, and stomach contents are taken for toxicologic analysis.

A neutron activation test was obtained at the beginning of the autopsy (the hands are bagged at the scene).

Other evidence collected includes the recovered bullets, hair and nail samples, clothes, and recovered particles.

All evidence was collected by the autopsy technician and placed in an appropriately labeled envelope with the name of the deceased, to be submitted to the County Crime Lab.

Microscopic Examination

The microscopic examination is consistent with the gross findings and final pathological diagnoses.

Endnotes

1. Steven D. Ashley, *Managing Training: Recapturing Lost Opportunities* (http://www.sashley.com/articles).
2. Craig Floyd, A record of Law Enforcement Sacrifice during the Twentieth Century, National Law Enforcement Memorial Fund Chairman Report, January 6, 2000.
3. Ibid.
4. *Brown v. City of Pinellas Park*, 557 So.2d 161, 172 (1990).
5. Wisconsin Department of Justice, Law Enforcement Standards Board, Pursuit Guidelines, 1995.
6. Resolution submitted by the IACP Highway Safety Committee at the IACP Annual Conference held in Phoenix, Arizona, 1996.
7. IACP "Sample" Vehicular Pursuit Policy dated October 30, 1996.
8. Geoffrey Alpert, *Police Pursuit: Policies and Training*, National Institute of Justice, Research Brief, May 1997.

Appendix: IACP Vehicular Pursuit Policy

Effective Date: October 30, 1996
Subject: Vehicular Pursuit CALEA Standard Ref: 41.2.2, 61.3.4
Reevaluation Date: October 30, 1999

I. Purpose
   The purpose of this policy is to establish guidelines for making decisions with regard to vehicular pursuit.

II. Policy
   Vehicular pursuit of fleeing suspects can present a danger to the lives of the public, officers, and suspects involved in the pursuit. It is the responsibility of the agency to assist officers in the safe performance of their duties. To fulfill these obligations, it shall be the policy of this law enforcement agency to regulate the manner in which vehicular pursuits are undertaken and performed.

III. Definitions
   A. Vehicular Pursuit: An active attempt by an officer in an authorized emergency vehicle to apprehend a fleeing suspect who is actively attempting to elude the police.
   B. Authorized Emergency Vehicle: A vehicle of this agency equipped with operable emergency equipment as designated by state law.
   C. Primary Unit: The police unit which initiates a pursuit or any unit which assumes control of the pursuit.
   D. Secondary Unit: Any police vehicle which becomes involved as a backup to the primary unit and follows the primary unit at a safe distance.

IV. Procedures
   A. Initiation of Pursuit:
      a. The decision to initiate pursuit must be based on the pursuing officer's conclusion that the immediate danger to the officer and the public created by the pursuit is less than the immediate or potential danger to the public should the suspect remain at large.
      b. Any law enforcement officer in an authorized emergency vehicle may initiate a vehicular pursuit when the suspect exhibits the intention to avoid apprehension by refusing to stop when properly directed to do so. Pursuit may also be justified if the officer reasonably believes that the suspect, if allowed to flee, would present a danger to human life or cause serious injury.
      c. In deciding whether to initiate pursuit, the officer shall take into consideration:

- Road, weather, and environmental conditions;
- Population density and vehicular and pedestrian traffic;
- The relative performance capabilities of the pursuit vehicle and the vehicle being pursued;
- The seriousness of the offense; and
- The presence of other persons in the police vehicle.

B. Pursuit Operations:

   a. All emergency vehicle operations shall be conducted in strict conformity with applicable traffic laws and regulations.

   b. Upon engaging in a pursuit, the pursuing vehicle shall activate appropriate warning equipment.

   c. Upon engaging in pursuit, the officer shall notify communications of the location, direction and speed of the pursuit, the description of the pursued vehicle, and the initial purpose of the stop. The officer shall keep communications updated on the pursuit. Communications personnel shall notify any available supervisor of the pursuit, clear the radio channel of non-emergency traffic, and relay necessary information to other officers and jurisdictions.

   d. When engaged in pursuit, officers shall not drive with reckless disregard for the safety of other road users.

   e. Unless circumstances dictate otherwise, a pursuit shall consist of no more than two police vehicles, a primary and a secondary unit. All other personnel shall stay clear of the pursuit unless instructed to participate by a supervisor.

   f. The primary pursuit unit shall become secondary when the fleeing vehicle comes under air surveillance or when another unit has been assigned primary responsibility.

C. Supervisory Responsibilities:

   a. When made aware of a vehicular pursuit, the appropriate supervisor shall monitor incoming information, coordinate and direct activities as needed to ensure that proper procedures are used, and shall have the discretion to terminate the pursuit.

   b. Where possible, a supervisory officer shall respond to the location where a vehicle has been stopped following a pursuit.

D. Pursuit Tactics:

   a. Officers shall not normally follow the pursuit on parallel streets unless authorized by a supervisor or when it is possible to conduct such an operation without unreasonable hazard to other vehicular or pedestrian traffic.

   b. When feasible, available patrol units having the most prominent markings and emergency lights shall be used to pursue, particularly as the primary unit. When a pursuit is initiated by other than a marked patrol unit, such unit shall disengage when a marked unit becomes available.

c. Motorcycles may be used for pursuit in exigent circumstances and when weather and related conditions allow. They shall disengage when support from marked patrol units becomes available.

d. All intervention tactics short of deadly force such as spike strips, low speed tactical intervention techniques, and low speed channeling (with appropriate advance warning) should be used when it is possible to do so in safety and when the officers utilizing them have received appropriate training in their use.

e. Decisions to discharge firearms at or from a moving vehicle, or to use roadblocks, shall be governed by this agency's use of force policy, and are prohibited if they present an unreasonable risk to others. They should first be authorized, whenever possible, by a supervisor.

f. Once the pursued vehicle is stopped, officers shall utilize appropriate officer safety tactics and shall be aware of the necessity to utilize only reasonable and necessary force to take suspects into custody.

E. Termination of the Pursuit:

a. The primary pursuing unit shall continually re-evaluate and assess the pursuit situation including all of the initiating factors and terminate the pursuit whenever he or she reasonably believes the risks associated with continued pursuit are greater than the public safety benefit of making an immediate apprehension.

b. The pursuit may be terminated by the primary pursuit unit at any time.

c. A supervisor may order the termination of a pursuit at any time.

d. A pursuit may be terminated if the suspect's identity has been determined, immediate apprehension is not necessary to protect the public or officers, and apprehension at a later time is feasible.

F. Interjurisdictional Pursuits:

a. The pursuing officer shall notify communications when it is likely that a pursuit will continue into a neighboring jurisdiction or across the county or state line.

b. Pursuit into a bordering state shall conform with the law of both states and any applicable inter-jurisdictional agreements.

     c. When a pursuit enters this jurisdiction, the action of officers shall be governed by the policy of the officers' own agency, specific inter-local agreements and state law as applicable.

G. After-Action Reporting:

     a. Whenever an officer engages in a pursuit, the officer shall file a written report on the appropriate form detailing the circumstances. This report shall be critiqued by the appropriate supervisor or supervisors to determine if policy has been complied with and to detect and correct any training deficiencies.

     b. The department shall periodically analyze police pursuit activity and identify any additions, deletions or modifications warranted in departmental pursuit procedures.

H. Training:

Officers who drive police vehicles shall be given initial and periodic update training in the agency's pursuit policy and in safe driving tactics.

Note: This sample policy is intended to serve as a guide for the police executive who is interested in formulating a written procedure to govern vehicular pursuit. IACP recognizes that staffing, equipment, legal, and geographical considerations and contemporary community standards vary greatly among jurisdictions, and that no single policy will be appropriate for every jurisdiction. We have, however, attempted to outline the most critical factors that should be present in every pursuit policy, including the need for training, guidelines for initiating and terminating pursuits, the regulation of pursuit tactics, supervisory review or intervention, and reporting and critique of all pursuits.

(Approved at the 103rd IACP Annual Conference, Phoenix, Arizona, October 30, 1996.)

Excited Delirium

5

Excited delirium is a term that is being used to explain the deaths of individuals who have been involved in a struggle with the police. It is controversial term because there has been no formal recognition of the phenomenon by the medical community, and it is not recognized in the *Diagnostic and Statistical Manual of Mental Disorders*. Even though the American Medical Association does not recognize this diagnosis as a medical or psychiatric condition, the National Association of Medical Examiners has recognized it for more than a decade. It is used by medical examiners in most major cities. Thus, there is a great deal of controversy regarding the use of this syndrome to explain sudden death while restrained.[1] However, the one thing that the medical community does agree upon is that excited delirium is a "medical emergency" no matter what the cause.[2] The law enforcement community believes that excited delirium is a physical condition that sometimes leads to death after a physical struggle and also believes that the key to avoiding these deaths is to train law enforcement officers on the recognition of the symptoms of excited delirium.

The *Canadian Medical Association Journal* has dismissed excited delirium as a "pop culture phenomenon." The Royal Canadian Mounted Police (RCMP) hired Compliance Strategy Group to author an independent study of excited delirium. In their report, the authors stated that the concept of excited delirium should be removed from RCMP training manual, policies, and procedures.

TASER® International has funded numerous studies to prove that excited delirium is a valid cause of death in cases where the excessive use of force and the use of TASERs were shown to have been involved.

Proponents of excited delirium as a legitimate diagnosis say that it usually strikes people who use large amounts of stimulants, especially cocaine or methamphetamines, and the mentally ill. In 1849, Luther V. Bell, M.D., the superintendent of Massachusetts McLean Asylum for the Insane, published a description of what appears to be the first case of excited delirium. Since that time, what today is called excited delirium has been variously known as Bell's Mania, agitated delirium, excited delirium, and acute exhaustive mania. Most of the early papers describing the condition speak of a prolonged period of increasingly bizarre behavior, usually over several days or weeks. This bizarre period seems to be much shorter when the victim is abusing stimulants such as cocaine or methamphetamines. Officers tend to see the bizarre and alarming behavior of a subject experiencing excited delirium as strictly a control-and-arrest situation rather than a serious medical emergency that can turn fatal. Proponents of excited delirium as a legitimate medical condition say that just as it became important for police officers to distinguish between a combative drunk and a person having a diabetic crisis, there needs to be training so that officers can recognize and distinguish between people choosing to act in a violent and criminal way, and those who are doing so because of an underlying medical condition that is affecting them both mentally and physically.

Opponents of excited delirium theory say they have never seen any proof that someone can be excited to death. The American Civil Liberties Union (ACLU) and the National Association for the Advancement of Colored People (NAACP) fear that the condition is being exploited and used as a medical scapegoat for police abuse. They believe most of these people do not die from drugs or some mysterious syndrome but from confrontation, abuse, and inappropriate use of force and restraint during a violent encounter that should have been avoided. They theorize that the cause is due to the psychological stress of being confronted with aggression that results in further physiological reactions (e.g., adrenaline release, increased heart rate, temperature, and strength), leading to death. The fact that many of these deaths happen during or soon after restraint clearly implies police abuse. The ACLU believes that most in-custody deaths are the result of excessive force and improper restraint techniques such as hog-tying and the use of pepper spray.[1]

Excited delirium has become recognized as a police-related cause of death that can be experienced in any size community, where it may be suddenly confronted by a patrol officer who is ill prepared to either recognize this medical emergency or know how to best respond. Although the phenomenon has been known to exist for many years, it was rarely encountered until more recently, and the greater occurrence is due to certain societal developments:

- Public policy has discouraged institutionalization of the mentally ill, while approximately 5% of the U.S. population, many of whom are homeless, suffer from a serious mental illness.
- Street abuse of cocaine and methamphetamine (meth) is increasing.

Not surprisingly, the primary candidates for being potential victims of excited delirium include:

- The mentally ill (primarily bipolar or paranoid schizophrenic), who are not institutionalized and are off their medications.
- Street drug abusers (primarily cocaine or crystal meth).

- Those who are suffering from both a combination of mental illness and street drug abuse.

Although the actual physiology of excited delirium is not yet fully understood, medical examiners are more frequently listing it as being at least a contributory cause of death.

Because unscientific and speculative media reporting of such deaths has caused widespread public criticism of in-custody deaths, police trainers have started to educate themselves about excited delirium and have started to develop training programs that are designed to prepare officers and minimize potential excited delirium deaths.

What Would You Have Done?

This incident takes place in the suburb of a metropolitan area that is served by a medium-sized police department of 150 officers and acts totally independent of the larger city's staff and training or communications facilities.

Sean Richards was 26 years of age and had been diagnosed as a paranoid schizophrenic since high school. He had been generally responsive to therapy and successfully medicated for several years, but unfortunately, and as is often the case, his medications had always produced unwanted side effects. His new girlfriend, Jennifer, had recently introduced him to crystal meth, and Sean soon found that he could "feel good" without the unwanted side effects of his psychotropic medications. Within a few weeks, he had completely quit taking his prescribed medications and was heavily into crystal meth, which he and Jennifer were able to manufacture in one room of the small home that they rented in a high-crime neighborhood of the central city.

One particularly hot summer afternoon, Sean and Jennifer were high and decided to supplement their income by a shoplifting foray into the neighboring suburb's shopping mall. While riding the bus to their destination, Jennifer noticed that Sean was becoming agitated, but he reassured her that he was okay and just worried about being caught by mall security. After leaving the bus and while walking to the mall entrance, Sean began sweating profusely and started mumbling to himself. Without warning, he suddenly began shouting, "The devil is trying to kill me," and started tearing off his clothes until he was completely naked. The entrance to the mall displayed a large water fountain that cascaded onto plate glass panels, and Sean leaped into the water, where he began bashing his head against the glass until he was bleeding profusely. Although his speech was mostly incoherent, he could sometimes be understood to yell, "I'm God, you can't kill God!" While shoppers looked on in disbelief and apprehension, a woman called 911 on her cell phone and reported to the suburban police dispatcher, "Some naked guy is in the mall fountain shouting that he's God." The dispatcher obtained the caller's identifying information and dispatched the nearest unit to the scene.

A solo officer soon arrived, parked with his emergency lights still activated, and ran up to the fountain. This officer was a large young man who lifted weights, was physically fit, and had a reputation for being able to handle himself in any physical confrontation. To the contrary, Sean was short, skinny, and obviously drunk or high. This should be no problem, the officer thought, and immediately asserted his command and control presence. Although Jennifer had been tempted to run, she knew that something was seriously wrong with Sean, whom she loved in her way, and she tried to talk with the officer. She wanted to at least tell him that Sean was a diagnosed paranoid schizophrenic, was "off his meds," and maybe even admit that he was high on crystal meth, too. She never had a chance before the officer told her, "Get back and stay out of this." The officer next shouted at Sean, "Get out of there, you're under arrest," and that is when Sean's attention turned to the officer. Growling like an animal, he waded toward the officer, climbed out of the fountain, and charged forward screaming, "You can't kill God!" The officer was totally unprepared for the attack and suddenly found himself unexpectedly struggling with a person who displayed superhuman strength and felt no pain. After managing to radio, "Officer needs help," in a clearly exhausted voice, he was able to access his oleoresin capsicum (OC) and covered Sean's face with spray but to no effect. The officer dropped his empty OC canister and extended his Asp baton, but repeated strikes also had no effect. Likewise, repeated drive stuns from his TASER had no effect. He was in a fight with Superman and had just concluded that he would have no alternative but to shoot in self-defense, when the first backup unit arrived, and a second officer joined the fight. Sean, however, threw both officers around with seeming ease, and they could not even begin to restrain him. The third and fourth units arrived next, and they, too, joined the fight, each trying to control an arm or leg, as they finally managed to place Sean face down on the ground. A sergeant was the fifth to arrive, and by the mutual efforts of all five officers, they managed to attach a handcuff to each wrist and join them together behind his back. Even with that amount of restraint, Sean managed to continue his struggle and was kicking wildly, as he repeatedly arched his body off of the ground. A sixth officer arrived with a hobble, which was wrapped around Sean's crossed ankles and secured to his handcuffs, but he still struggled with maximum physical exertion. Accordingly, the two latest arriving officers held him down with the weight of pressure on his back and buttocks, while the others caught their breath.

It seemed like the more they tried to hold him down, the more he struggled until he suddenly "calmed down" and gave the last two officers an opportunity to catch their breath, too. After a few minutes and while waiting for medics to arrive, one of the officers noticed that their prisoner did not seem to be breathing, and a quick check of his carotid pulse found none. The restraints were removed, he was rolled over onto his back, and cardiopulmonary resuscitation (CPR) was being started as the medics arrived to take charge of the patient. Although the medics managed to start his heart and breathing again en route to the hospital, he never regained consciousness.

The nightly news reported that a mentally ill man was severely beaten by the police, hog-tied, and died in police custody. The story was repeated in the local press, and an editorial demanded that the officers be held criminally responsible.

Best Practices Based upon What We Know

- There are estimated to be between 50 and 125 in-custody deaths each year in the United States that correlate with excited delirium symptoms.
- There is no medical or psychiatric diagnosis of excited delirium.
- The International Association of Chiefs of Police (IACP) does not acknowledge the syndrome.
- Each year more medical examiners blame excited delirium for in-custody deaths.
- Excited delirium symptoms include:
  - Imperviousness to pain
  - Great strength
  - Hyperthermia
  - Sweating
  - Bizarre and violent behavior
  - Aggression
  - Hyperactivity
  - Hallucinations
  - Confusion and disorientation
  - Foaming at the mouth
  - Drooling
  - Dilated pupils
- It is accepted in the law enforcement profession that excited delirium is a medical emergency that requires acute medical care.
- It is accepted in the law enforcement profession that officers must be trained on recognition of excited delirium.
- It is accepted in the law enforcement profession that law enforcement agencies must have an established protocol on dealing with subjects in excited delirium.

- If a TASER is used, it is recommended that one firing in probe mode is all that should be used, as multiple activations increase the risk of death.
- Officers should be trained so they are aware that a high risk of sudden death is associated with people in excited delirium.

Simply stated, officers confronted with a situation in which the suspect is believed to be in a state of excited delirium should immediately call for backup officers, should immediately call for emergency medical personnel to be ready to treat the suspect as soon as he or she is under physical control, and should use a TASER in electromuscular disruption mode one time and attempt restraint while the suspect is incapacitated by the TASER. If the TASER is ineffective, the officers will have little choice but to go hands-on with the suspect to gain physical control and restraint.

Investigation of the Scene

In these types of situations, officers should communicate with headquarters immediately to report the incident and request an ambulance and additional backup personnel be dispatched right away. The shift supervisors or commanding officer should respond to the scene immediately. The following tasks should be implemented:

- Assist and remove any injured individuals from the scene as soon as possible.
- Secure the crime scene.
- Begin crime scene protection measures.
- Use barrier tape, official vehicles, or necessary measures as needed.
- Set up a command post in accordance with departmental operations policy.
- Notify the appropriate personnel in the departmental chain of command to take charge of the situation.
- Employ crowd control procedures to protect the integrity of the scene.
- Detain witnesses—witnesses have valuable information about the incident.
- Try to separate all witnesses to prevent discussions about the incident.
- To avoid a potential confrontation between the officer or officers and the victim's family, remove the officer or officers involved from the scene.
- Initiate crime scene investigation procedures.
- Notify the appropriate agencies, including the forensic laboratory, medical examiner, and other law enforcement agencies when required.
- Notify the decedent's family and community leaders when required.
- Release the department's official statement related to the incident.

General Crime Scene Procedure

The general crime scene procedure should be followed. Major crime investigators, crime scene technicians, and forensic specialists should be called to the scene. Scientists and investigators should work together methodically and swiftly to process the scene. The purpose of crime scene investigation is to document and memorialize the scene, to collect and preserve forensic evidence, and to reconstruct the incident event.

The basic elements of a crime scene investigation include the following:

1. Crime Scene Survey
   The lead investigator and case officer should conduct a quick scene survey and determine the crime scene parameter. Avoid any unnecessary delay.
2. Documentation of the Crime Scene
   The crime scene should be documented according to standard procedures, and a four-corner crossover photographing method should be used for document the area.
3. Crime Scene Search
   There are seven commonly used methods for a crime scene search. Figure 5.1 illustrates these methods. Line search, strip method, and grid method are used for outdoor scenes. Link searches are used for indoor scenes. Spiral and wheel methods and zone search are for special situations.

   Crime scene search patterns are varied and outwardly different in style and application. However, they all share a common goal of

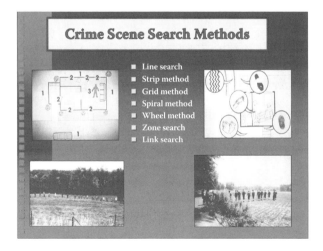

Figure 5.1 Methods of crime scene search patterns. (**See color insert following page 78.**)

providing structure and organization to ensure that no physical or pattern evidence is overlooked. There is no single correct method for a specific type of crime scene.

Line or Strip Methods

This approach involves the demarcation of the crime scene into a series of lines or strips. Then, members of the search team are arranged at regular intervals, usually at arm's length, and then proceed to search along straight lines. The investigator identifies any evidence in his or her path. This method is also referred to as the strip method. The crime scene coordinator regulates the pace of the search. This method is well suited for searching large areas, such as parks, fields, yards, parking lots, or highways. In addition, it can be conducted by as few as one or two investigators or as many as hundreds.

Crime scenes that involve multiple gunshots must be searched for all related firearms evidence, shell casings, and projectiles. An inventory of the weapons and bullet magazines potentially involved in the incident will identify the maximum amount of cartridges and bullets that may be found at the scene. The information can be supplemented with a preliminary examination of the deceased's wounds. If basic search methods do not recover all of the firearms evidence, then an organized line method search should be conducted. In an investigation concerning a police officer–related shooting when the initial scene search was unable to locate one of the expended shell casings, the scene search should include the use of metal detectors. Locating the shell casing is important, as it is information needed to assist in the reconstruction process conducted months later.

Grid Method

The grid method is a modified double-line search. In this approach, a line pattern is constructed, and then a second line pattern is established in the same area, running perpendicular to the first line pattern. Searchers follow the first line pattern and search in the same manner as with the line method. Upon completing the first line pattern, the searchers realign on the other line pattern. Thus, the same area is searched twice by a grid pattern format. An additional advantage is that two different searchers search the same area. Although this method is more time consuming, it has the advantage of being more thorough and methodical.

Wheel Method

With the wheel method, the crime scene is considered to be essentially circular. The investigators start from a critical point and travel outward along many straight lines, or rays, from this point. This search pattern becomes increasingly difficult when searching larger areas and, therefore, is usually used only for special scene situations and with limited applications.

Spiral Method

Similar to the wheel/ray method, the spiral method considers the crime scene as circular in design. There are two techniques commonly used for spiral methods: one method is generally referred to as an inward spiral and the other is an outward spiral. With the inward spiral method, investigators start at the outer boundary and circle the crime scene toward the critical point. With the completion of each circuit, the diameter of the circle is progressively decreased until a central point is reached. Meanwhile, the outward method involves starting at the critical point and circling outward. These approaches rely on the ability to trace a regular pattern of fixed diameter. Thus, physical barriers at the scene may pose problems during the search. The spiral method is generally used for special conditions of the crime scene search. However, there is a danger that physical evidence could be destroyed while walking to the central point to initiate the search. A very limited number of situations require the application of these search methods.

Zone Search

Crime scenes that include readily definable zones can be effectively searched by focusing on the zones in a systematic manner. Indoor crime scenes are examples of such a scene. Depending on the size of the scene, each zone may be subdivided as needed until it is of manageable size. There are a variety of techniques for conducting a zone pattern search. If the search is to be conducted by a small group of trained crime scene investigators, the entire team can work together in a particular zone. For example, if the scene involved a private residence, the investigators could enter one room at a time, document, search, and collect all relevant evidence as a unified team. For the primary areas of the scene, this methodology is recommended. However, if there are numerous zones in ancillary areas, then the team can be divided, and each member can conduct a search of a particular zone. If this method is chosen, it is good practice to have a particular zone searched twice by two different investigators. The scope of the search and the type of evidence being sought will help determine the appropriate method to search in a zone format. Zone searching is also advantageous in that the individual zone can be prioritized. Critical zones such as target areas, point of entry, and point of exit can be searched multiple times.

Link Search

The link method is often the most productive and common approach for crime scene searches. This method is based on the four-way linkage theory, seeking to find associations between the scene, victim, suspect, and physical evidence. With this method, the investigators evaluate the scene and then proceed through the area in a systematic and logical fashion to gather physical evidence that can be linked or associated to a particular crime or activity.

Collection, Preservation, and Packaging of Physical Evidence

In this particular scenario, the following types of physical evidence should be collected:

1. From the Police Officer Involved in the Shooting
   a. Weapon, if firearm was discharged
   b. Ammunition, if firearm was discharged
   c. Clothing, shoes, tear and damage patterns
   d. Gunshot residue (GSR) kit of hands, if firearm was discharged
   e. Injury pattern
   f. Bloodstain
   g. OC can
   h. Soil, grass, stains
   i. Other relevant evidence
2. From Decedent
   a. Clothing and shoes
   b. Bullet or bullets and fragments, if firearm was discharged
   c. Bullet trajectory, if firearm was discharged
   d. GSR kit and pattern on clothing if firearm was discharged
   e. Trace evidence
   f. Weapon (knife) if possessed by suspect
   g. OC residues on clothing and body
   h. Drug, containers, documents on body
   i. Complete toxicology results
3. Evidence at Scene
   a. Spent casings and their locations if firearm was discharged
   b. Spent bullets and their locations if firearm was discharged
   c. Weapon and its location
   d. Bloodstains and their patterns
   e. Body location
   f. Scuff marks and pattern evidence
   g. Conditional evidence

    h. Trace and transfer evidence
    i. Bullet trajectory if firearm was discharged
    j. Location and condition of TASER
    k. Location and condition of OC can
    l. Location and condition of baton
    m. Any vomited material or other body fluids

Preliminary Reconstruction

Preliminary crime scene reconstruction should be conducted to determine the sequence of events and the location of each subscene within the overall crime scene.

Releasing the Scene

Once the scene processing is complete, the scene should be released quickly. All the debris, bloodstains, markings, and crime scene tape should be removed from the area.

TASER and Excited Delirium

Some persons involved in an altercation with law enforcement agents die as a result of an altercation when TASERs are involved. Some studies suggest the reason for many of these deaths is based on the fact that the person fighting with a law enforcement officer is suffering from a controversial condition known as *excited delirium*. Excited delirium is said to occur when a person has a state of mind "begetting irrational and often violent behavior." It has been noted that persons suffering from excited delirium often had been using cocaine at the time of the altercation. Cocaine is a drug that causes heart arrhythmias when ingested by a person. The victim's drug use may be unknown to the police officers at the time of the altercation. As already stated, there is much debate surrounding excited delirium as a legitimate medical condition, and as the stated cause of death in altercations involving the use of a TASER on an out-of-control and noncompliant victim.

Law enforcement agents depend on nondeadly weapon alternatives such as TASERs to safely subdue suspects and defuse potentially dangerous situations. Recently, serious injuries and deaths have been linked to TASER use and have prompted law enforcement agencies and the public to question the weapon's safety.

Many police suspects have been injured or killed after being shocked with TASERs. Officers have been injured during training. Critics believe the TASER should be used only as an alternative to lethal force, and that law enforcement agents currently use the device too freely. These critics cite a rash of recent injuries caused by the device on cadets of law enforcement agencies, suspects not imposing an imminent danger to themselves or others, and even on uncooperative children, the elderly, and pregnant women.

From 2001 to 2007, there were over 250 deaths associated with police use of TASERs throughout the United States and Canada. The number of TASER-related deaths has increased by the year as more law enforcement agencies have included TASERs as standard-issue weapons to their officers. One noted organization reports the death toll by TASER use in the United States and Canada as follows: 1 in 1999, 1 in 2000, 4 in 2001, 13 in 2002, 19 in 2003, 56 in 2004, 67 in 2005, and 69 in 2006. This statistic alone demonstrates that as the number of TASERs in the hands of law enforcement agencies grows, so does the number of deaths of TASER victims.

A 2005 British Columbia study examined current information regarding law enforcement agents' use of TASERS and the impact of the electrical shocks on subjects. The study found that persons shocked by TASERs might suffer serious bodily harm or even death after being shocked by the supposedly nonlethal device. TASER International officially recognizes only a few health risks associated with its weapons: direct injuries from the impact of TASERs include abrasions, scarring, and eye damage from the laser aim. Strong muscle contractions may also cause injuries that may include the following: hernias; ruptures; internal injuries to soft tissues, organs, muscles, tendons, ligaments, nerves, and joints; and bone fractures including fractures to vertebrae. TASER International also cautions about secondary injuries caused by the inevitable loss of control and fall of the stunned victim. Those who are physically infirm, pregnant, or located on unstable or elevated platforms may be at a higher risk of secondary injury after receiving a TASER shot.

Other sources indicate a number of significant health risks associated with TASER use that are not recognized by TASER International. For example, a shock from a TASER can occur during the vulnerable period of a heartbeat cycle. This vulnerable period is the stage in a heartbeat cycle during which an electroshock is highly likely to cause ventricular fibrillation. Ventricular fibrillation is a state in which the heart muscles spasm uncontrollably, disrupting the heart's pumping function and causing death. Certain segments of the population may be more susceptible to ventricular fibrillation, such as children, due to their small size and, in direct relation, the small size of their hearts. Also more susceptible to ventricular fibrillation are individuals on psychiatric medication or other drugs due to the effect of such drugs of raising blood pressure in the body. Multiple applications of TASER shocks can also naturally increase the chance that the electrical charge will hit the heart during the vulnerable period.

Other Cases Involving Excited Delirium

On February 13, 2006, Darvel Smith struggled with police after leaving a New Orleans bar. Smith was tased by Louisiana State Police during the struggle. After being tased, Smith was placed in handcuffs. Shortly after the TASER blast to his back, Darvel Smith died of cardiac arrest. A pathologist for the Orleans Parish Coroner's Office stated, "Darvel Smith died as a result of excited delirium and the presence of cocaine in his system." The coroner went on to state that the TASER stun did not contribute to Smith's death. Smith's family sued for wrongful death, but the case was dismissed because the force used by the officers was reasonable under the *Graham* test.

Police responded to a disturbance on the evening of April 16, 2006. Billy Ray Cook had been seen running through the streets screaming "Please don't shoot me!" Cook was incoherent, delusional, and under the influence of impairing substances. Three officers attempted to handcuff Cook, but Cook resisted and was tased numerous times until the officers were able to handcuff and shackle him. Between the three officers, Cook was tased a total of 38 times. The chief medical examiner determined the main cause of death in the case to be excited delirium due to cocaine intoxication. The medical examiner did state that Cook's altercation with the officers "may have been a contributing proximate cause of his death." Cook's estate sued the county and the officers for wrongful death, but summary judgment was granted for the defendants due to the fact that the officer's use of force was reasonable under the *Graham* test.

Sergio Galvan died of a heart attack after a struggle with police officers on March 23, 2007. Galvan was in a residential neighborhood, not far from his home, and was screaming incoherently. Galvan was approached by two police officers and began fighting with them. Galvan was tased during the struggle and died shortly thereafter. The medical examiner's report indicated four sets of lesions, which were interpreted as being marks left by the TASER. The coroner listed Galvan's cause of death as a result of "Excited Delirium Syndrome due to acute intoxication with cocaine." A wrongful death suit filed by Galvan's estate ended in summary judgment for the defendants, the City of San Antonio, and the officers involved in the struggle that resulted in Galvan's death. The court found the force applied to Galvan to be reasonable under the *Graham* test set forth by the U.S. Supreme Court.

Phillip Wayne LeBlanc died April 1, 2004, after being briefly detained by the Los Angeles Police Department. Police officers came upon LeBlanc that evening and instantly believed that he was under the influence or mentally ill. LeBlanc was handcuffed to a fence, and at that time "his speech became unintelligible growling." LeBlanc was sweating profusely even though it was a cool night, and he seemed impervious to pain. The police rejected using pepper spray or a beanbag gun to control LeBlanc, and instead agreed the best course of action would be to tase him. LeBlanc was tased twice and restrained by the officers. LeBlanc stopped breathing while he was restrained and soon after died. The coroner determined the cause of death to be excited delirium caused by cocaine intoxication. The coroner went on to claim that the TASER discharge did not contribute to LeBlanc's death.

LeBlanc's estate sued the city for wrongful death and had four expert witnesses testify to dispute the excited delirium diagnosis. Dr. Gary Ordog testified that the amount of alcohol and cocaine in LeBlanc's system at the time of his death was nonlethal, and that the cause of death was cardiac arrest resulting from the TASER discharge. Dr. Dennis Hooper testified that the cocaine found in LeBlanc's blood was nonlethal, and that the TASER charges deployed by the officers were "the major contributing cause of death." Dr. Michael Morse testified that the use of a TASER by the police in the circumstances surrounding the case should be deemed a use of deadly force. Finally, Gary Clark, a 27-year veteran of the Los Angeles County Sheriff's Department, testified that the officers involved did not allow sufficient time for the situation to deescalate, and that the use of a TASER in that circumstance was tactically improper. Even with the plaintiff's expert testimony calling the coroner's diagnosis directly into question, the court granted summary judgment for the defendants.

Model Legislation for the Safe Use of TASERS

Recognizing the abundance of TASER-related deaths and health risks, states should recognize the need for regulating the use of TASERS by law enforcement agencies. Only New Jersey outright forbids TASER use, which is too drastic of a measure because TASER use is beneficial in many situations that require nonlethal force. Successful TASER legislation should include standard reporting and training requirements for all law enforcement officers.

Massachusetts has a well-developed TASER statute in regards to reporting requirements that should be in place in model TASER safety legislation. The reporting requirements under the Massachusetts statute require the assaulting officer to report the number of times the weapon has been fired as well as to identify the characteristics of the individual at whom it was fired. This information is likely already found in police reports, so the reporting requirements would not be burdensome extra paperwork for officers.

TASER International's X26 TASER model is capable of recording the time, date, and duration of each discharge. There is a data port on the X26 model that connects through a USB cable to any computer for ease of recording the data. The X26 model also has an optional TASER Cam, which records both audio and video of the TASER discharge. The TASER X26 would be an ideal choice for law enforcement agencies attempting to comply with the reporting requirements, because most, if not all, of the needed data is saved on the TASER.

The raw data collected by the law enforcement agencies should then be transmitted to an independent clinic, such as in a university, for annual analysis. The results would then be forwarded to government officials and agencies to review and to determine if any remedy is necessary. This practice is similar to that required in Massachusetts.

The model legislation for regulating TASERS should also mirror the Massachusetts statutes in regard to training requirements. Massachusetts requires TASER training programs to be developed using guidelines set forth by a government agency and requires that the programs be approved by that agency before implementation. Specific issues regarding health risks should be covered extensively in the training programs. Other issues such as minimum shock duration, recognition of intoxicated or excited persons, medical protocol, and location on use-of-force continuum should all be covered in the mandatory training program.

Finally, the model TASER statute should include strict use requirements similar to those found in Florida's statute. The language of the Florida statute states that situations when police may deploy TASERS "must involve an arrest or a custodial situation during which the person who is the subject of the arrest or custody escalates resistance to the officer from passive physical restraint to active physical resistance," and the suspect must be "preparing or attempting to flee or escape." Requirements such as these would prevent the use of TASERs against those suspects who pose no risk to the officer or to themselves.

Endnotes

1. Mary Paquette, Excited Delirium: Does It Exist? *Perspectives in Psychiatric Care*, September 2009.
2. C.W. Lawrence and J.T. Cairns, Sudden Custody Death: The Ontario Perspective, *RCMP Gazette*, 2001.

Suicide-by-Cop (SbC) Incidents

6

Any shooting deliberately induced by a subject (intentionally forcing a confrontation with an officer which leaves the officer no choice but to use deadly force) is classified as a suicide-by-cop (SbC).[1]

According to a report released in 1983 of 99 officer-involved shootings (OISs) in Los Angeles County by Dr. Karl Harris, former Los Angeles County Deputy Medical Examiner, between 10% and 25% of all the shootings he studied involved suicide attempts.[2] Harris also found that 50% of the weapons used to threaten officers in the SbC incidents he studied were firearms. Because the overwhelming majority of the firearms were operative and loaded, it can reasonably be concluded that SbC incidents can be very hazardous to the responding officers.

Another study by Constable Rick Parent of the Delta, British Columbia, Police Department in 1996 looked at OISs by municipal police and Royal Canadian Mounted Police and found similar results with 10% to 15% of the cases being premeditated suicides.[3]

The most recent study of SbC incidents, conducted by Kris Mohandie, J. Reid Meloy, and Peter I. Collins, was reported in the *Journal of Forensic Sciences* in March 2009. The study examined 707 cases of North American OISs from 1998 to 2006. The study found that 36% of OISs were SbCs. In 50% of the incidents, the outcome was the death or injury of the subject.[4]

As was found in the Harris study, other studies have confirmed that these incidents can be very hazardous to the officers responding. The most recent found that 80% of the subjects possessed a weapon. In 60% of the incidents, the weapon was a firearm. Of those having a firearm, 50% discharged the firearm at the police during the encounter, and 19% simulated weapon possession to induce the police to shoot them.[5]

Indicators of Suicide-by-Cop

Some who have studied the SbC phenomenon have identified basic indicators that qualify an incident as a SbC. One such study established four criteria:

1. The subject verbalized a wish to die and asked the police to kill him or her.
2. The subject left a suicide note or verbalized a desire to commit suicide to a family member or friend.

3. The subject possessed a lethal weapon or what could reasonably appear to be a lethal weapon.
4. The subject intentionally escalated the incident or provoked the officers into shooting him or her.[6]

Barry Perrou, founder of the Public Safety Research Institute, identified 15 indicators that can help an officer recognize when he or she may be facing a SbC situation:

1. The subject is barricaded and refuses to negotiate.
2. The subject has just killed someone, particularly a close relative, his mother, wife, or child.
3. The subject says that he or she has a life-threatening illness.
4. The subject's demands of the police do not include negotiations for escape or freedom.
5. The subject has undergone one or more traumatic life changes (death of a loved one, divorce, financial devastation, etc.).
6. Prior to the encounter, the subject gave away all of his or her possessions and money.
7. The subject has a record of assaults.
8. The subject says he or she will only surrender to the person in charge.
9. The subject indicates that he or she has thought about planning his or her death.
10. The subject expresses an interest in wanting to die in a "macho" way.
11. The subject expresses an interest in "going out in a big way."
12. The subject expresses feelings of hopelessness.
13. The subject dictates his or her will to negotiators.
14. The subject demands to be killed.
15. The subject sets a deadline to be killed.[7]

One of the most difficult aspects involved in studying the SbC phenomenon is that researchers must generally rely upon incidents of OISs which resulted in the death of the subject and then attempt to determine if the death was a SbC. However, barricaded subject incidents, or incidents involving responses to the mentally ill or emotionally disturbed persons that are successfully resolved often are not reported on in newspapers. As a consequence, many actual incidents of SbC are not recognized and recorded as SbCs.

A Typical SbC Scenario

This incident takes place in a rural county served by a small sheriff's office of only 19 deputies, and there is a local state police detachment nearby, which provides mutual aid upon request.

Ralph Anderson was a young man, but he had a troubled marriage and was no stranger to the local sheriff's office, which had responded to the Anderson residence four times on domestic violence calls during the past year.

The domestic violence incidents had been increasing in severity (as is the usual pattern), and the last two calls had involved physical violence that resulted in Ralph's arrest. In both cases, he was released from jail on the following morning and went home to his wife, who accepted him back in the home, because he was the sole means of support for her and her two small children.

Ralph finished work at his auto repair job early on Friday and decided to have a few beers with a buddy at a local tavern. As time passed and more beer was consumed, Ralph left the tavern just before midnight and managed to drive home DUI without being apprehended. He awakened his sleeping wife and wanted to have sex, but she wanted nothing to do with her drunken husband. He forced himself upon her anyway, after which he fell asleep and she called 911 to report spousal rape. When the deputies arrived, they arrested Ralph and booked him into the county jail on the felony charge, but before leaving the home, one deputy took the wife aside. He advised her that she had best make arrangements to protect herself this time, because her husband would likely soon be released and might angrily retaliate against her. The deputy also asked her if she had a safe place to stay and told her how to obtain an order of protection against her husband. Fortunately, her parents lived nearby, where she could stay with them, and she obtained an order of protection on Monday morning, which was filed with the sheriff's office and served on Ralph, who was still in jail. Ralph was arraigned on the spousal rape charge on Monday afternoon and released on bond, whereupon he immediately went home to find that his wife had taken the children and moved out. He also found a receipt for his guns, which had been seized by the sheriff's office after his arrest for spousal rape, and the issuance of an order of protection. He suspected that she had gone to her parents' home, but he was afraid of her father, a retired rural district firefighter, who had friends in the sheriff's office.

Feeling generally put upon, abandoned by his family, and depressed by the accumulated events of the last four days, Ralph began to drink, while he pondered his troubles. Becoming increasingly despondent, his thoughts turned to suicide, but without his guns, he could not imagine how to kill himself until he had the thought, "Why not let the deputies do it for me?" Accordingly, he dialed 911 and, in a drunken slur, told the call taker/dispatcher, "I just cut the f---ing bitch's head off and, if you come out here, I'll f---ing kill you, too!" The dispatcher immediately broadcast that information to the only two deputies on duty and gave them the address, from which the 911 call had been received. In the meantime, Ralph had another drink and waited to see what would happen next. He did not have to wait long before he heard sirens and saw flashing lights, as the first unit closely followed by the second arrived in his driveway.

The lead deputy announced their presence over his public address system and ordered any occupant(s) to come out of the house. Now that the reality of impending death was suddenly upon him, however, Ralph was not sure what he actually wanted to do but armed himself with a large butcher knife and remained behind the door without answering. Repeated orders to come outside went unanswered, and the deputies conferred. They decided to have their dispatcher call the home, and although they could hear the phone ringing inside, it was not answered. After nearly half an hour had passed, they approached the house and tried to see inside through several unobstructed windows, but nothing seemed out of place and the suspect remained hidden. As more time passed and nothing further was known, the deputies decided they had exigent circumstances to enter the home, and they approached the front door, which they found to be unlocked.

As the contact deputy slowly pushed the door open, from the small front porch on which he was standing, his partner covered him from below. Suddenly and forcefully, the door was pulled wide open and the suspect instantly appeared holding an upraised knife in his right hand. The contact deputy jumped back from the threat, just as his partner fired in surprise, and that first shot struck the contact deputy in the lower back, causing him to continue falling backward off the porch. Having heard the sound of his own shot and seeing his partner fall to the ground, the cover deputy fired multiple rounds of what he believed to be return fire into the suspect.

The fatally wounded suspect had successfully committed SbC, the wounded deputy survived to retire as a paraplegic, and the shooting deputy retired on a stress-related disability.

What Would You Have Done?

Using the above scenario, identify those characteristics of Ralph Anderson that made it likely that the deputy was facing a subject intent on SbC. Identify the tactics that you would have employed in the above scenario. Identify the tactics you believe should be avoided in the above scenario. Were there any opportunities for surprise and the use of nonlethal or less-lethal weapons?

SbC Resolution Tactics

For the officer who must deal with a person who has barricaded himself or herself or is acting in an abnormal or bizarre manner and is threatening to take his or her own life, the task of determining what action to take is complex. It seems reasonable that the starting point must be a quick evaluation by the officer of the subject's behavior in the total context of the situation to determine whether the indicators of a SbC incident are present.

Next, aggravating factors must be considered. Is the subject armed with a firearm? Is the subject armed with a knife? If unarmed, are weapons available to the subject? Does the subject appear to be under the influence of alcohol or a drug? Has the subject committed a crime? Is the subject holding a hostage? Each one of those factors dictates which tact, or combination of tactics, might result in a successful resolution.

Next, information must be sought and assessed, such as to what degree of reaction does the officer get when attempting to communicate with the subject? Are family members, friends, or neighbors available who can give information as to what may be the subject's intentions?

Next, the officer must evaluate the circumstances to determine whether the subject can be surprised with nonlethal force and overpowered. Or, are the circumstances such that less-lethal options should be attempted?

It seems obvious that the more information learned by the responding officer, the greater the possibilities are for a successful resolution (no harm to the officer or the subject). Unfortunately, this all takes time, and the subject most often controls the element of time, and equally important, the subject who is armed with a firearm forces the officer into an officer-safety-first mode that means communication with the subject from a position of cover is demanded as the first step toward resolution.

Tactical recommendations for officers responding to possible SbC incidents have been made. One example is the Student Reference Manual for Suicide by Cop Incidents produced by the California Commission on Peace Officer Standards and Training in 1999. The manual emphasizes recognition of the type of person likely to be involved in a SbC incident and awareness of the danger posed to the officer. The manual advises officers to avoid a confrontation if possible, to stall for time, to use trained negotiators, and to consider the use of less-lethal force. The recommendations were very consistent with the recommendations made by the National Law Enforcement Policy Center in its Concepts and Issues Paper on dealing with the mentally ill person published in 1997.[8]

Vivian B. Lord studied 64 SbC incidents in North Carolina and found that the primary tactical strategy utilized by the law enforcement officers involved was negotiations. In 28 cases, negotiation attempts resulted in some discussion of the problem with the subject, and 93% were resolved successfully. In seven cases, gas, as a less-lethal effort, was utilized, and 88% of those cases were successfully resolved.[9]

If negotiation tactics are employed, negotiators should be alert for the indicators that suggest intervention is working. Barry Perrou identified several of those indicators, including:

- Less interactive tension
- Lowered voice
- Less anger
- Less profanity
- Diminished aggressive body language
- Increased nonaggressive body language
- Diminished threats of violence
- Less hopelessness and helplessness
- A greater willingness to listen to officers' suggestions[10]

Investigation at the SbC Scene

The supervisor responding to an officer shooting scene is often the foundation for the successful resolution of the case. The basic duties of the first-responding supervisor (FRS) to the scene of a suspected SbC incident are to:

1. Assist the injured individuals: Officers, victims, and witnesses should all be treated equally.
2. Crowd control: To protect the integrity of the scene and evidence.
3. Detain witnesses: Witnesses have valuable information about the incident; try to separate the witnesses to prevent discussions about the incident; and of course, the witnesses will need to be interviewed.
4. Protect the crime scene: Begin the crime scene protection measures. Use barrier tape, official vehicles, or other measures as needed.
5. Communicate with supervisors.
6. Remove officers involved in the shooting from the scene to avoid potential confrontation between officers involved and family members. In addition, their weapon and ammunition should be collected and tagged along with clothing. A gunshot residue (GSR) kit should also be collected.
7. Initiate crime scene investigation procedures.
8. Notification of the incident: In accordance with departmental guidelines or practices, notify the appropriate personnel in the departmental chain of command, and notify the forensic laboratory medical examiner and other appropriate agencies as required; notify the prosecutor's office when required.

9. Notification of the incident to decedent's family
10. Media Information: Release department official statement related to the shooting incident. The name of the officer(s) should not be released at this point.

General SbC Scene Procedures

Crime scene investigation is a discovery process. The investigation seeks to discover all the aspects of the shooting incident and all the activities at the scene. This includes such facts as the nature of the shooting, the variety of physical evidence associated with the scene, medical history of the decedent, and possibly, any sequencing of the decedent's behavior prior to shooting.

The scene must be methodically and systematically examined. This examination will include many steps that must be done properly and have to follow a certain sequence; the omission of a step could possibly cause valuable pieces of evidence to be overlooked or to not meet the legal or scientific requirements.

Conducting the SbC Crime Scene Survey

Once the scene is secure, the crime scene supervisor or lead investigator with the case officer should conduct a preliminary crime scene survey.

The following is a suggested guideline listing of the tasks that should be performed during the crime scene "walk-through":

1. Note all the essentials, such as time, location, lighting, and position and condition of potential evidence.
2. Mentally begin a *preliminary* reconstruction of the events that might have led to the facts of the shooting. This step is not the act of forming a rigid or fixed theory of how the shooting occurred; it is only a beginning point of the reconstruction. Keep in mind that scientific crime scene investigation is objective and systematic. To allow it to be otherwise risks developing "tunnel vision" and may mislead the direction of the investigation.
3. Note the types of transient and conditional evidence present at the scene. Be aware of weather conditions (and their changeability), light switches on or off, door locks intact or broken, windows opened or closed, heating, ventilation or air conditioning status, presence of odors, location of physical evidence, and so forth. At this point, documentation, protection, preservation, and collection of these special forms of physical evidence should be considered.

Figure 6.1 Crime scene investigator thoroughly documenting the scene. **(See color insert following page 78.)**

4. Be aware of the legal implications for crime scene searches. Any search of a person, house, location, or vehicle should meet the requirements established by law.

Documentation of the Crime Scene

Any crime scene must be thoroughly documented or recorded by notes, photography, sketching, and videotaping. Sometimes the use of audiotaping is a useful means as part of the documentation process (see Figure 6.1).

Crime Scene Search

Crime scene searches involve both the surrounding area and the target area of a scene. The primary purpose of a crime scene search is the recognition and collection of physical evidence. Unless the potential physical evidence can be recognized and collected, there is no forensic testing that can be conducted.

Collection, Preservation, and Packaging of Physical Evidence

There are many types of physical evidence found at a crime scene. Figure 6.2 shows the general categories of physical evidence commonly found at a scene.

Biological	**Chemical**	**Physical (Impression)**
Blood	Fibers	Fingerprints
Semen	Chemicals	Firearms
Saliva	Glass	Handwriting
Body Fluids	Soil	Printing
Hair	Gunpowder	Number Restoration
Botanical	Metal	Footprints
Bone	Mineral	Tire Marks
Tissues	Narcotics	Tool Marks
Urine	Drugs	Typewriting
	Paper	
	Ink	

Figure 6.2 Common types of physical evidence found at a crime scene.

Assignment of Responsibility

One investigator should be assigned as the physical evidence collector. This individual will have the responsibility to collect, mark, preserve, and package the evidence found at the scene. In this particular shooting scenario, the following types of physical evidence should be collected:

1. From police officer(s) involved in shooting
   a. Weapon
   b. Ammunition
   c. Clothing, shoes
   d. GSR kit of hands
   e. Injuries
   f. Bloodstains
   g. Other relevant evidence
2. From decedent(s)
   a. Clothing and shoes
   b. Bullet(s) and fragments
   c. Bullet trajectory
   d. GSR kit and pattern on clothing
   e. Trace evidence
   f. Weapon (knife, gun, etc.)
3. Evidence at scene
   a. Spent casings and their location
   b. Spent bullets and their location

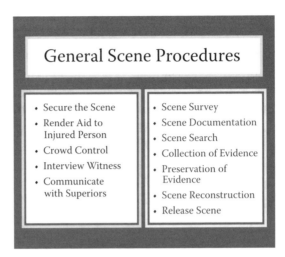

Figure 6.3 General crime scene procedures for the first responder (left) and the crime scene investigator (right). **(See color insert.)**

   c. Weapon and its location
   d. Bloodstains and their patterns
   e. Body location
   f. Pattern evidence
   g. Conditional evidence
   h. Trace and transfer evidence
   i. Bullet trajectory

Preliminary Reconstruction

Crime scene reconstruction involves scene information analysis, interpretation of scene patterns, and examination of the physical evidence (Figure 6.3). The crime scene should be viewed and photographed from several different angles with all possible alternatives. The scene as it exists, the physical evidence and its examination results, and the use of simple logic, both deductive and inductive, along with common sense, will tell what happened, when, where, who with, and how it happened.

Releasing the Scene

The crime scene is released only when it is reasonably certain that all physical evidence was collected and answers were obtained.

Summary of Crime Scene Procedures

Suicide-by-Cop, Autopsy Report

External Examination

The body is that of a well-developed, adequately nourished African American male, weighing 198 pounds, measuring 77 inches, and appearing to be stated age of 23 years.

The body is dressed in the following articles of wearing apparel: a brown/tan/green sweat jacket, a navy blue T-shirt with the logo's name printed on the front, knee-length blue jean shorts with a black belt, white and blue boxer shorts, white socks, tennis shoes, and brown, cotton work gloves. The jacket and T-shirt show damage: five, longitudinally oriented, vertical, irregular-linear tears are noted along the posterolateral seam of the jacket. A round hole is noted to be present on the left posterior lower quadrant of the back of the jacket. Three, irregular-oval tears are situated to the right of the posterior portion of the seam of the left sleeve. Two groups of three and two round holes are noted to be present horizontally on the left posterolateral aspect of the T-shirt. These lesions match the gunshot wounds on the left posterior thorax and left shoulder, which will be described below.

No jewelry, rings, or watch are present.

The temperature of the body is cool to the touch. Rigor mortis is well developed and present to an equal extent in all joints. Purple, poorly visible due to natural skin pigmentation, nonfixed livor mortis is evident over the posterior parts of the body. The body shows no evidence of decomposition. The skin is smooth. A few, small fragments of broken glass are present on the body.

The head and face exhibit trauma, which will be described below. The face is randomly smeared with blood. The head hair is black and short. The eyes are brown with pale conjunctivae. The corneas and lenses are transparent. No petechial hemorrhages or congestion are noted in either conjunctiva. The pupils are regular, round, equal, central, and measure 0.5 cm in diameter. The ears and external auditory canals are unremarkable. The skeleton of the nose is intact, and no foreign material is present in the nostrils. A small colorless, tied plastic bag with white, solid material is found in the oral cavity beneath the tongue. The bag is placed in a properly labeled envelope with the name of the deceased and given to the autopsy technician for transfer to the crime laboratory. The gums are pink and normal. The upper and lower teeth are natural and in a good state of dental repair. The lips, oral mucosa, and tongue reveal no evidence of trauma.

The neck and the shoulders are symmetrical. The left shoulder shows a gunshot wound, which will be described below.

The chest is symmetrical and exhibits injuries, which will be described below.

The abdomen is flat and no masses can be palpated through the abdominal wall.

The back is symmetrical and exhibits trauma, which will be described below.

The external genitalia and the anus are unremarkable. The penis is circumcised, and the testes are present in the scrotum.

The extremities are symmetrical. The fingernails are irregular, clean, and very short. The toenails are clean, short, and unremarkable. The skin of the legs exhibits no changes. No edema is present in the ankles or legs.

Passive motion of the neck, shoulders, elbows, wrists, fingers, hips, knees, and ankles fails to elicit any bony crepitus or abnormal motion.

No fresh needle marks or punctiform scars are noted in either antecubital fossa, interphalangeal spaces of the hands or feet, under the tongue or on the gums.

Evidence of Recent Trauma

There are three gunshot wounds on the body. They will be described below in descending anatomic order.

    I. Gunshot Wound of Head; Bullet Recovered
      A. Gunshot wound of entrance:

At the medial end of the left eyebrow and immediately below it, 12.5 cm below the level of the top of the head and 2 cm to the left of the midline of the face, there is an irregular-oval gunshot wound of entrance with loss of skin, measuring 2 × 1 cm, with two narrow connective tissue tags that are protruding outward from the aforementioned defect. The margins of the entry wound are hemorrhagic with minor lacerations, (0.1 to 0.2 cm in length), radiating outward from the main defect. An irregular, 0.1- to 0.2-cm wide, dark red rim of abrasion is noted along the entire free margin of the wound. A small fragment of a copper bullet jacket is embedded in the subcutaneous tissue of the gunshot wound of entrance. Several pink-red, well-marginated, pinpoint abrasions are situated on the left half of the forehead on the dorsum of the nose and in both zygomatic regions, occupying an area measuring 10 × 11 cm. Soot is not detected around this gunshot wound of entrance. Multiple, superficial nicks and linear abrasions are clustered on the forehead and in the area of the left eyebrow.

      B. Track of the bullet:

The bullet perforates the left supraorbital margin of the frontal bone, greater wing of the sphenoid bone, and then into the cranial cavity. Then the projectile perforates the meninges and transects the pons, totally destroying this area. There is a mild, dark red, hemorrhage along the pathway of the bullet.

C.   Site of recovery of the bullet:

A lead bullet, markedly deformed, and a copper bullet jacket have been retrieved with gloved fingers, without instruments, from the destructed pons. The parts of the bullet are placed in a properly labeled envelope with the name of the deceased and given to the autopsy technician for transfer to the County Crime Lab.

D.   Trajectory of the bullet:

The trajectory of the bullet is backward and slightly downward.

II. Gunshot Wound of the Left Posterior Thorax; Bullet Recovered

A.   Gunshot wound of entrance:

On the posterior aspect of the thorax, 42.5 cm below the level of the top of the head and 22 cm to the left of the posterior midline of the body, there is a round, regular gunshot wound of entrance with loss of skin, measuring 0.6 cm in diameter. It is surrounded by a dark red, dry, and slightly indented abrasion collarette, measuring 0.2 cm in width. No soot or powder is noted around or outside this gunshot wound entrance.

B.   Track of the bullet:

The bullet perforates the entire thickness of the left thoracic wall, penetrates into the left pleural cavity, and penetrates the distal portion of the upper lobe of the left lung. Then, the bullet proceeds upward, lacerating the upper lobe of the left lung longitudinally, and creating a wide, markedly hemorrhagic track in the parenchyma of the upper lobe. Then the bullet exits the left lung, creating a large hemorrhagic laceration near the apex. The bullet ascends further, penetrating into the soft tissues of the left anterior neck and creating a nick of the left internal carotid artery, causing a large infiltrating hemorrhage in the soft tissues around the arteries and nerves on the left anterior medial neck. There is a marked, dark red hemorrhage along the entire track of the bullet. There is 1700 mL of dark red, bloody fluid present in the left pleural cavity.

C.   Site of recovery of the bullet:

A deformed copper jacketed, lead bullet is retrieved with gloved fingers from the space immediately to the left of the left internal carotid artery, 5.5 cm above the level of the clavicle. The bullet is placed in a properly labeled envelope with the name of the deceased and given to autopsy technician for transfer to the County Crime Lab.

  D. Trajectory of the bullet:

   The trajectory of the bullet is forward, upward, and rightward.

 III. Gunshot Wound of the Left Shoulder, Through-and-Through

  A. Gunshot wound of entrance:

   On the posterolateral aspect of the left shoulder, 31 cm below the level of the top of the head, there is a round, regular gunshot wound of entrance with loss of skin, measuring 0.5 cm in diameter. It is surrounded by a regular, red-brown, slightly indented rim of abrasion, 0.2 cm wide. No soot or powder is noted around or inside this gunshot wound of entrance.

  B. Track of the bullet:

   The bullet perforates the skin and subcutaneous tissue only and proceeds leftward and downward, creating a slightly hemorrhagic track, 5.5 cm in length, in the subcutaneous space only.

  C. Gunshot wound of exit:

   On the posterolateral aspect of the proximal third of the left upper arm, 35 cm below the level of the top of the head, there is an oval gunshot of exit, measuring 0.6 × 0.4 cm. The distal end of this wound exhibits a pink, 0.9 × 0.4 cm indented abrasion.

Other Identifying Features

Multiple, small, superficial scars are randomly located on the upper and lower limbs.

Internal Examination

Body Cavities The body is opened by a "Y"-shaped incision. The abdominal fat pad is 1 cm thick at the umbilicus. The undamaged muscles of the chest and abdominal wall are normal in color and consistency. The ribs, sternum, and spine exhibit no fractures. The gunshot wounds of the left thorax and the left hemothorax were previously described. The right pleural cavity is smooth and moist. The peritoneal cavity is moist. The liver and spleen do not extend below the costal margins. The bladder lies below the symphysis pubis. The organs of the pleural and peritoneal cavities are in their usual positions in situ. The mesentery and omentum are unremarkable.

Neck The skin of the neck, the thyroid and cricoid cartilages, larynx, and the hyoid bone show no evidence of traumatic injury. There is a marked hemorrhage in the soft subcutaneous tissues of the left anteromedial neck, which was previously described (site of recovery of the bullet). Posterior dissection of the neck reveals no traumatic injuries. The laryngeal mucosa is pink. The epiglottis and vocal cords are unremarkable.

Cardiovascular System   The heart weighs 400 grams. The pericardium contains a few cc of clear liquid. The epicardial surface is smooth. The external configuration of the heart is unremarkable. The right and left ventricles are unremarkable. The endocardium and valve leaflets are smooth and transparent and exhibit no thrombi, vegetations, or fibrosis. The trabeculae carneae and papillary muscles are unremarkable. The chordae tendineae are usual. The right ventricle is 0.3 cm thick, and the left ventricle is 1.5 cm thick. The coronary arteries have their usual distribution with a right predominance. The coronary ostia are normal in patency. Multiple cross sections at 0.2-cm intervals show no pathological changes. The myocardium is of usual consistency, dark brown, and grossly homogeneous.

The aorta is unremarkable.

The venae cavae are unremarkable.

Respiratory System   The right lung weighs 475 grams, and the left lung weighs 400 grams. The through-and-through gunshot wound of the left lung was previously described. The tracheal mucosa is unremarkable. The pleurae are delicate and glistening. The lungs are distended and are variegated pink-gray to red-gray. The lung parenchyma is of the usual consistency and mottled with a minimal amount of anthracotic pigment. The lung tissue is not congested. No nodularity is seen.

The extra- and intrapulmonary bronchi contain blood. The pulmonary arteries and veins exhibit no pathological change. The hilar and mediastinal lymph nodes are not enlarged.

Hepatobiliary System   The liver weighs 1595 grams. The capsule of Glisson is transparent. The external surface is smooth, glistening, and brown. The borders are sharp. The parenchyma is of usual consistency and brown with the usual lobular architecture and no focal or diffuse lesions.

The gallbladder has delicate walls, and contains 0.5 to 1 cc of bile. It has a smooth mucosa. No stones are present.

The intra- and extrahepatic biliary ducts are patent. The hepatic and portal veins and the hepatic artery are unremarkable.

Hemolymphatic System   The spleen weighs 160 grams and has a normal consistency. The capsule is glistening and thin. The internal architecture is clearly defined.

Gastrointestinal System   The esophagus is empty and unremarkable. The stomach contains 200 cc of undigested food (corn and other vegetables). The remainder of the gastrointestinal system is unremarkable.

The appendix is identified.

Urogenital System   The right kidney weighs 160 grams, and the left kidney weighs 150 grams. The surfaces are smooth and glistening. The capsules strip easily, revealing smooth, red-brown surfaces. The corticomedullary junctions are well defined. The calyceal and collecting systems are not remarkable. The renal arteries and veins are unremarkable.

The ureters are not dilated or obstructed.

The bladder contains 200 cc of clear, yellowish urine. The bladder exhibits the usual mucosa and muscularis. The ureteral orifices are patent.

The prostate is not enlarged and does not impinge upon the urethra. The tissue of the prostate is lobulated, tan and moderately firm.

Endocrine System   The adrenals, thyroid, parathyroids, pancreas, and pituitary are not remarkable.

Musculoskeletal System   The muscles are well developed and show the usual color and consistency. The sternum, ribs, and spine exhibit the usual bone density and marrow.

Central Nervous System   The scalp is reflected, revealing trauma. The gunshot wound of the head was previously described. The calvarium is removed, revealing evidence of minimal epidural and subdural hemorrhages. The dura mater does not exhibit any stains or discolorations. The leptomeninges are slightly and diffusely hemorrhagic.

The brain weighs 1695 grams and is of usual consistency. The sulci occupy their usual positions, and the gyri are flattened. The blood vessels at the base do not reveal any aneurysms. The cerebral and cerebellar hemispheres are symmetrical, and the surface does not display any scar tissue. The ventricles contain a small amount of bloodstained cerebrospinal fluid. The cerebellar tonsils are herniated. Multiple sections through the cerebrum, cerebellum, pons, midbrain, and medulla exhibit the usual internal pattern. The pons have been totally destroyed.

The skull shows an irregular, oval fracture in the area of the sphenoid bone (left) with small, linear fractures radiating backward.

Note: Blood, urine, eye fluid, and stomach content are taken for toxicologic analyses.

A neutron activation test was obtained at the beginning of the autopsy.

Other evidence collected includes the projectiles and hair.

All evidence was collected by the autopsy technician and placed in appropriately labeled envelopes with the name of the deceased, to be submitted to the County Crime Lab.

Microscopic Examination
The microscopic examination is consistent with the gross findings and final pathological diagnoses.

Anatomic Diagnoses

I. Gunshot Wound of Face and Head; Bullet Recovered
   A. Gunshot wound of entrance—area of left eyebrow
   B. Fractures of skull
   C. Destruction of pons
   D. Intracranial hemorrhages, minor
   E. Site of recovery of bullet—pons
   F. Trajectory of bullet—backward and slightly downward

II. Gunshot Wound of Thorax, Posterior, Left; Bullet Recovered
   A. Gunshot wound of entrance—left posterior thorax
   B. Perforation of thoracic wall
   C. Perforation of upper lobe, left lung, through-and-through
   D. Hemothorax, left, marked, 1700 mL
   E. Penetration into soft tissues of left anterior neck
   F. Bullet recovered—near the left internal carotid artery
   G. Trajectory of bullet—forward, upward, and slightly rightward

III. Gunshot Wound of Left Shoulder, Through-and-Through, Superficial
   A. Gunshot wound of entrance—left posterior shoulder
   B. Track of bullet—subcutaneous space only
   C. Gunshot wound of exit—left posterior-lateral aspect of left upper arm
   D. Trajectory of bullet—leftward, forward, and slightly downward

Opinion

A 19-year-old African American male died as a result of gunshot wounds of the thorax and head. A through-and-through superficial gunshot wound of the left shoulder is also noted.

Manner of Death

The manner of death is homicide.

The Future

As more effective less-lethal weapons become available to law enforcement, it is likely that more SbC cases will be resolved successfully. Conducted energy devices seem to have the greatest potential for immediate incapacitation with minimal harm to the subject in incidents in which weapons other than a firearm are being used by the subject. In any case, as more SbC cases are identified and the tactics used are analyzed, it is not unrealistic to think that law enforcement officers in the future will be more successful in dealing with SbC subjects.

Endnotes

1. International Association of Chiefs of Police (IACP) Training Keys 535 and 536 entitled *Officer Assisted Suicide*, 2001, page 37.
2. Studies by Dr. Karl Harris, former Deputy Medical Examiner, Los Angeles County, California, 1983.
3. Richard B. Parent, *Victim-Precipitated Homicide: Aspects of Police Use of Deadly Force in British Columbia*, Simon Fraser University, British Columbia, Canada, July 1996.
4. Kris Mohhandie, J. Reid Meloy, and Peter I. Collins, Suicide by Cop among Officer-Involved Shooting Cases, *Journal of Forensic Sciences* 54(2), 456–462, 2009.
5. Ibid.
6. H. Hutson, D. Anglin, J. Yarbrough, K. Hardaway, M. Russell, J. Strote, M. Canter, and B. Blum, Suicide-by-Cop, *Annals of Emergency Medicine* 32(6), 665–669, 1998.
7. Scott Buhrmaster, *15 Warning Signs That You Might Be Involved,* PoliceOne.com News, April 7, 2006 (http://www.policeone.com/pc_page2).
8. IACP National Law Enforcement Policy Center, *Dealing with the Mentally Ill,* Concepts and Issues Paper, December 1, 1997.
9. V.B. Lord, *Law Enforcement-Assisted Suicide: Characteristics of Subjects and Law Enforcement Intervention Techniques,* New York, Looseleaf Law Publishers, 2001.
10. Vivian B. Lord, *Suicide by Cop-Inducing Officers to Shoot: Practical Direction for Recognition, Resolution and Recovery*, New York, Looseleaf Law Publishers, 2004.

Positional Asphyxiation 7

Positional asphyxia, also known as restraint asphyxia, is the cessation of adequate breathing by means of restraint and can occur by either positioning to compromise the airway, compression to inhibit the respiratory function, or a combination of both such mechanisms. Although some have denied the contribution of these breathing inhibitors to in-custody death and claim that their scientific testing excludes any possibility of that causation, they admittedly have conducted their tests under laboratory conditions that bear no resemblance to the realities of the street, where these deaths continue to occur.

What Would You Have Done?

This incident takes place in a suburb of a large city, where the local police department had recently taken a zero-tolerance position against condoning domestic violence by law enforcement officers. Officer Arthur Thomas, who worked on the large city's police department, was in his fourth year on the job and loved it, but, as is not infrequently the case, his marriage was experiencing difficulties. His young wife Rita was pregnant with their first child and feeling increasingly lonely, with her parents living over in the next state and her husband working the graveyard shift, which required him to sleep during the day. Accordingly, she would often wait up for Officer Thomas to come home after shift, but as a means to wind down from a busy night, he would increasingly have an after-shift beer or two with his fellow officers. One early morning incident, where shots had been fired at the police officers responding to an armed robbery in progress, was especially stressful, and Officer Thomas, along with several of his buddies, had more to drink than the nominal one or two beers, and he came home much later than usual. Nevertheless, Rita had waited up for him but had become increasingly irritated with his late arrival and greeted him at the door with, "Where have you been and how much have you had to drink?" Unfortunately, Officer Thomas unthinking and too quickly replied, "I have been with the only friends I have and it is none of your business how much I drink!"

Wrong answer, and the exchange began to escalate. The people in the apartment beneath the Thomas family, who did not know them, heard the sounds of a scuffle over shouted threats and called 911 to report an ongoing domestic violence incident. Any police officer who has responded to a domestic violence call at 5:00 A.M., knows that it is not going to be fun, and this was no exception. The responding male and female officer could hear the loud dispute as soon as they arrived, and they quickly went to the source's door, where they knocked loudly to demand entrance. Rita Thomas opened the door crying, and upon seeing the two local officers, she warned them that her husband had a gun. They responded by pulling her outside, slammed the door behind her, took cover down the hallway, and called for backup. After their sergeant arrived and they had time to question Rita further, they found that her husband was a police officer in the nearby large city and dispatch was asked to call Officer Thomas outside with a warning to not be armed. He cooperated and was soon talking to the two suburban officers and their sergeant in the hallway outside of his apartment.

As the responding officers discussed the domestic violence call with the young couple, it was learned that Officer Thomas had shoved his wife during their argument. Accordingly, the local sergeant advised Officer Thomas that he was required to place him under arrest, seize his firearm, and advise the police department that employed him of the whole matter. Officer Thomas knew that this would probably end his cherished police career, and he implored the local officers to cut him a break. They explained that they could not grant his request, and with his judgment probably affected by both alcohol and his seemingly desperate circumstances, he tried to go back into his apartment, and the fight was on. While the sergeant and his two officers struggled with Officer Thomas, additional backup was called. Two more officers immediately arrived on the scene from where they had been staging nearby. With a combination of an adrenaline pump and his defensive tactics training in ground fighting, Officer Thomas was more than holding his own, until he was given a three-second burst of oleoresin capsicum (OC) spray in his face and was wrestled to a prone position on the floor.

Handcuffs were forcibly applied behind his back, and because he continued to squirm and struggle, his ankles were hobbled too, after which his feet were drawn up toward the handcuffs and they were attached together, in what is commonly called the *hog-tied* position. He continued to struggle against his restraints, however, so while the sergeant and first two officers caught their breath, the other two later arriving officers held him down to the floor. After a few minutes, one of the officers remarked, "I think he's playing possum," but when checked more closely, Officer Thomas was found to be not breathing, without a pulse, and unresponsive. The arresting officers quickly removed his restraints, rolled him on his back, and started cardiopulmonary resuscitation (CPR), while awaiting an advanced life support (ALS) unit to transport him to the hospital.

Although emergency medical treatment was able to start a heartbeat and breathing again, his brain was irreversibly damaged by lack of oxygen, and artificial life support was discontinued. His wife, the mother of their infant daughter, has sued the local police department.

Best Practices

Examination of positional asphyxia deaths has revealed that there are several factors that apparently play a role in a person's susceptibility to death. It has been noted that a large abdomen increases the risk of positional asphyxia, as such people when placed in a prone position suffer breathing difficulty because the contents of the stomach are forced upward in the abdominal cavity which puts pressure on the diaphragm, a critical muscle responsible for respiration. When this occurs, the diaphragm cannot move properly, and the person suffers breathing difficulty.

Another contributing factor is psychosis induced from the ingestion of alcohol and drugs which tends to be accompanied by outbursts of activity and loss of breath. Finally, many preexisting physical conditions can also contribute to a higher risk of death from positional asphyxia, as any physical condition that impairs breathing under normal circumstances (heart disease, asthma, emphysema, and bronchitis) tends to be exacerbated by exertion and restraint.[1]

Because of these deaths, law enforcement and health professionals have been taught to avoid restraining people in a prone (face-down) position and if necessary, to do so for very short time periods only. The International Association of Chiefs of Police (IACP) has addressed the risk of positional asphyxia in several training and policy guidance documents stating that prisoners should never be transported in a restrained and prone position due to the evidence suggesting that this can result in sudden deaths. Citing the work of noted pathologist Donald Reay and the fact that his work has been reaffirmed by other specialists, the IACP concluded that

> Positional asphyxia occurs when the position of the body interferes with respiration, resulting in asphyxia (e.g. suffocation) and this potential for death is exacerbated in situations where the suspect is overweight, has exerted substantial energy in resisting arrest or in other ways prior to being restrained, and when the suspect has ingested alcohol and/or drugs prior to the event.[2]

To best prevent actual or presumed cases of restraint asphyxia, the following are recommended:

1. Avoid prolonged struggles with multiple officers engaged on a sole subject.
   a. Slow events down and attempt to negotiate compliance and surrender.
   b. If not possible to avoid a physical struggle, be prepared to use overwhelming force at the earliest opportunity to avoid a prolonged and exhaustive physical confrontation.

    c. If a struggle begins and a suspect is placed in a prone position, the law enforcement officer should hold his or her breath. The moment the officer needs to breathe, the suspect needs to be rolled onto his or her side to prevent positional asphyxia.[3]

    d. The TASER is an excellent option for quickly gaining control.

2. Do not use the maximum restraint (*hog-tied*) position.

3. As soon as the subject is restrained, immediately and without delay

    a. Roll subject onto his or her side and sit upright.

    b. Monitor breathing and consciousness.

    c. Transport for emergency medical care.

Crime Scene and Forensic Evidence

In this scenario, police officers were dispatched to respond to a domestic violence situation. A sergeant and several backup officers arrived at the scene. After negotiations broke down, an arrest was made, and during the struggle, Officer Arthur Thomas died. Because no gun was discharged and no weapon was used, this type of scene has a very limited amount of physical evidence available. Fingerprints, hair, fibers, and even bloodstains have limited forensic value, because Officer Thomas lived in this apartment. However, other evidence such as broken furniture, type of clothing, and other pattern evidence could be valuable for reconstruction.

General Procedures

- Secure the apartment and the hallway area.
- Remove the decedent's wife from the scene. A community relations officer or an officer with special training should consult with the wife and remove her from the apartment.
- Remove the officers involved in the incident from the scene.
- Initiate crime scene investigation procedures.
- Identify all witnesses and take statements.
- Notify the appropriate agencies, including the crime scene specialist, the forensic laboratory, the medical examiner, and the district attorney's office.
- Release a department official statement related to the incident.

Indoor Scene Procedure

A general indoor crime scene search procedure should be followed. All the potential evidence should be documented and collected.

Figure 7.1 Forensic investigators examining an indoor scene. **(See color insert following page 78.)**

Figure 7.1 is a photograph of forensic investigators examining an indoor scene. Each piece of potential evidence should be carefully studied, documented, preserved, and collected for further laboratory examination. Pattern evidence might be crucial in the future of the investigation. Pattern evidence should be enhanced and documented for further analysis. Because this scene is limited to a confined area, the number of crime scene investigators allowed to enter the scene should be limited to avoid accidental destruction or alteration of the scene.

Crime Scene Survey

The crime scene survey should be conducted by the most experienced investigator. Location of bloodstains, scruff marks, broken objects, and damage to furniture should be recorded.

Documentation of the Crime Scene

The entire scene should be thoroughly documented using the standard crime scene documentation techniques. The following areas are especially important for reconstruction of the event:

- Photographs and measurement of all injuries
- Photographs and documentation of broken objects and furniture

- Photographs and documentation of all pattern evidence
- Photographs and documentation of all bloodstain patterns

Figures 7.2 and 7.3 depict many types of pattern evidence, such as blood-stains, broken glass, and damaged furniture found at the scene of a restraint asphyxia death.

Crime Scene Search

A combination of zone and link crime scene search methods should be used for this type of scene search.

Figure 7.2 Physical evidence located at a restraint asphyxia death scene. (**See color insert.**)

Figure 7.3 Trace evidence found at a crime scene. **(See color insert.)**

Collection, Preservation, and Packaging of Physical Evidence

The following types of forensic evidence need to be collected:

1. Evidence from the Police Officer Involved
   a. Injury or bruises
   b. OC container
   c. Clothing and shoes
   d. Statements related to incident
2. Evidence from Decedent
   a. Clothing and shoes, noting any tears, rips, or damages
   b. Bloodstain patterns
   c. Other pattern evidence on body
   d. Medical and autopsy reports
   e. Toxicology report
3. Evidence from Scene
   a. Bloodstains and patterns
   b. Other body fluids and patterns
   c. Body position and location
   d. OC can
   e. Broken objects
   f. Scuff marks on floors or walls
   g. Furniture damage
   h. Markings and other contact patterns

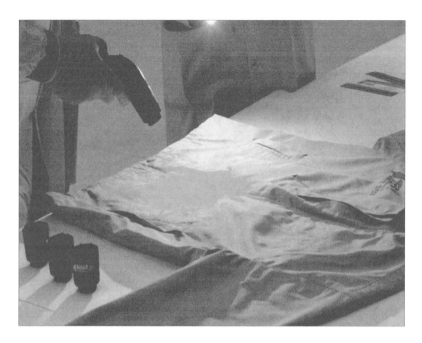

Figure 7.4 A laboratory scientist examines a shirt. **(See color insert.)**

   i. Bloody fingerprints, shoe prints, and other imprint evidence
   j. Alcohol or drug containers

Preliminary Reconstruction

One of the important aspects of this investigation is to reconstruct the event. Reconstruction is based on the forensic and pattern evidence found at the scene. Figure 7.4 shows a forensic scientist utilizing a forensic light source to examine tears, damage, and bloodstain evidence on the decedent's shirt.

Some of the pattern evidence is not visible under normal conditions. Various methods can be used to enhance those patterns. Figure 7.5 depicts enhanced bloody fingerprints and a footprint at a crime scene with blood enhancement reagent tetramethylbenzedine.

Autopsy Report 1—Positional Asphyxia Death

External Examination

The body is that of a well-developed, obese, African American male, weighing 330 pounds, measuring 72 inches, and appearing to be consistent with the age of 43 years.

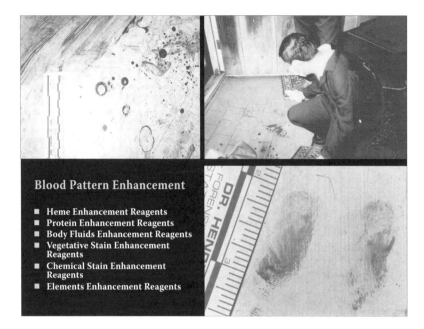

Figure 7.5 Blood enhancement reagent used to develop pattern evidence at a crime scene. (**See color insert.**)

The body is unembalmed and unclad.

No jewelry, rings, or watch are present.

The temperature of the body is cool to cold to the touch. Rigor mortis is well developed. Purple, nonfixed, marked livor mortis is evident over the posterior parts of the body, except in areas exposed to pressure, where it is absent. There is diffuse edema (i.e., anasarca). There is focal skin slippage of the right side of the back and around the posterior right wrist. The skin shows evidence of trauma, as described below. The head, neck, and upper chest are congested/show lividity.

The head is normocephalic.

The head and face exhibit trauma, which will be described below. The head hair is bald. The eyes are brown. The corneas and lenses are cloudy. Due to edema, the conjunctivae cannot be evaluated. The sclerae are also edematous with some hemorrhage in the left sclera. The ears and external auditory canals are unremarkable. The skeleton of the nose is intact. Bloody mucous is present in the nostrils. No foreign material is present in the oral cavity. The gums are normal. The upper and lower teeth are natural and in a good to fair state of dental repair. The lips, oral mucosa, and tongue reveal no evidence of trauma, though the tongue is clenched between the teeth. A mustache is present.

The neck is symmetrical and unremarkable. No increased mobility on manipulation is detected (Figure 7.6).

Figure 7.6 Officers engaged in a restraint asphyxia scenario. (**See color insert.**)

The shoulders are symmetrical.

The chest is symmetrical and unremarkable.

The abdomen is bulging, and no masses can be palpated through the abdominal wall. On the right lower quadrant is the outline of an electrocardiogram (EKG) patch with a 1.8 × 0.3 cm yellow abrasion in its superior border.

The back is symmetrical and exhibits trauma, as described below.

The external genitalia and the anus are unremarkable except the scrotum is swollen. The testes are palpable in the scrotum.

The extremities are symmetrical and exhibit trauma, which will be described below. The fingernails are dirty and short. The toenails are clean, short to medium in length, and several are dystrophic. The soles of the feet are calloused.

Manipulation of the neck, shoulders, elbows, wrists, fingers, hips, knees, and ankles fails to elicit any bony crepitus or abnormal motion.

The body shows the following evidence of trauma:

1. There is a 0.2 × 0.2 cm scabbed abrasion on the right side of the nose, toward the bridge.

2. There are two dry dark red/black abrasions on the upper back, measuring 0.8 × 0.2 cm (medial left side) and 1.8 × 0.5 cm (midline).
3. There is a 4.5 × 1 cm area of fairly linear confluent dry yellow-red to dark red abrasions on the medial right wrist, which could relate to handcuffs. An approximately 3 cm long and 0.1 cm wide curvilinear healing abrasion comes off the midportion of these linear abrasions, superiorly.
4. There are three dry pink-red abrasions on/around the left elbow, measuring 0.2 × 0.2 cm, 1 × 0.7 cm, and 2 × 1.7 cm, with the latter having dark red areas in it.
5. There are two dry dark red to yellow-red abrasions on the right knee, measuring 1.3 × 0.2 cm and 2.3 × 0.6 cm, with a 1 × 0.4 cm area of avulsed skin beneath the knee on the anterior right lower leg, proximally.

Note: X-rays of both hands and wrists did not reveal the presence of any fractures. Incisions are made into both sides of the back, from shoulders to buttocks and down to rib cage/scapulas, without evidence of any soft tissue hemorrhage.

Evidence of recent medical/surgical treatment:

1. There is an endotracheal tube in the oral cavity.
2. There is a nasogastric tube present, entering the left nostril.
3. There is bloody gauze in the right nostril.
4. There are intravascular catheters in the right antecubital fossa, anterior right wrist, and both femoral areas.
5. There is a needle puncture mark in the left antecubital fossa.
6. There is a Foley catheter present.

Other identifying features:

1. There is a multicolored tattoo on the anterior left upper chest.
2. There is a 17.5 × 0.5 cm vertical scar around the right side of the umbilicus and of the midline lower abdomen.
3. There is a 3.1 × 0.3 cm horizontal scar on the abdomen, in the right lower quadrant.
4. There are multiple hyperpigmented patches and scars across the upper abdomen/lower chest, individually measuring to 0.9 × 0.7 cm. On the left lower chest one of these may be a possible accessory nipple.
5. There are striae on both sides of the lower back.
6. There are several hyperpigmented patches/scars on the right elbow and posterior to medial forearm, individually measuring up to 7 × 1.8 cm.

7. There is a 2 × 0.4 cm scar on the anterolateral right upper arm, distally.

8. There are two scars on the dorsum of the left hand measuring 0.3 × 0.1 cm (with tiny scab present) and 0.7 × 0.7 cm.

9. There are multiple hyperpigmented patches/scars on both shins and knees, and focally on the distal anterior right thigh, individually measuring up to 3.2 × 2.6 cm. A 0.5 × 0.3 cm nearly healed abrasion on the left shin is in this area.

Internal Examination

Body Cavities

The body is opened by a "Y"-shaped incision. The abdominal fat pad is 2.5 cm thick at the umbilicus. The muscles of the chest and abdominal wall are normal in color and consistency with no evidence of soft tissue hemorrhage. The ribs, sternum, and spine exhibit no fractures. The domes of the diaphragm are normally positioned. The pleurae are smooth. Each pleural cavity contains approximately 700 cc of clear yellow fluid (i.e., bilateral pleural effusions). The peritoneum is smooth and thin. The peritoneal cavity contains approximately 300 cc of blood-tinged fluid (i.e., ascites). The blood seems to be secondary to postmortem liver temperature taking. The bladder lies below the symphysis pubis. The organs of the pleural and peritoneal cavities are in their usual positions in situ. The mesentery and omentum are unremarkable. The pulmonary artery is opened in situ, and no emboli are seen.

At this time, representative samples of blood, bile, and eye fluid are taken for toxicological examination.

Cardiovascular System

The heart weighs 640 grams. The pericardium is thin and smooth and contains 30 to 40 cc of clear yellow liquid (i.e., pericardial effusion). The epicardial surface is smooth. There is a moderate amount of epicardial fat. The external configuration of the heart is unremarkable. The chambers of the heart are of normal size. The left ventricle is hypertrophied. The endocardium and valve leaflets are smooth and transparent and exhibit no thrombi, vegetations, or fibrosis. The circumferences of the valves are as follows: tricuspid, 13 cm; pulmonic, 8.6 cm; mitral, 11 cm; and aortic, 7.3 cm. The trabeculae carneae and papillary muscles are unremarkable. The chordae tendineae are usual. The right ventricle is 0.3 to 0.5 cm thick, and the left ventricle is 1.6 to 1.7 cm thick. The septum is 1.8 and 1.9 cm thick. The coronary arteries have their usual distribution with a left predominance. The right and left coronary ostia are normal in patency. On sectioning, the coronary arteries display no significant degree of atherosclerosis or other pathological abnormality. The myocardium is of usual consistency, red-brown, and grossly homogeneous.

The aorta is lined by a smooth, tannish-yellow endothelium and shows fatty streaking.

The bifurcation of the iliacs is patent.

The venae cavae are unremarkable.

Respiratory System

The right lung weighs 990 grams, and the left lung weighs 900 grams. The tracheal and laryngeal mucosa are congested with patchy hemorrhage in them beneath the vocal cords and down most of the length of the trachea. The pleurae are delicate and glistening. The lungs are distended and are dark purple. The lung parenchyma is firm, rubbery, and all sections sink in formalin. The parenchyma is mottled with a moderate to marked amount of anthracotic pigment. The lung tissue is markedly congested and edematous. No purulent exudate is expressed from the parenchyma on compression. No nodularity and no focal or diffuse lesions are seen.

The extra and intrapulmonary bronchi are opened longitudinally, patent and congested. The pulmonary arteries and veins exhibit no pathological change. The hilar and mediastinal lymph nodes show anthracosis.

Hepatobiliary System

The liver weighs 3195 grams. The capsule of Glisson is transparent. The external surface is smooth and mottled light brown to red-brown. The borders are blunted. The parenchyma is congested and mottled light brown to red-brown.

The gallbladder has delicate walls, contains 30 cc of yellow-brown bile, and has a smooth mucosa. No stones are present.

The intra- and extrahepatic biliary ducts are patent. The hepatic and portal veins and the hepatic artery are unremarkable.

Hemolymphatic System

The spleen weighs 305 grams and is soft. The capsule is glistening and intact. The internal architecture is blurred due to congestion.

There are no enlarged lymph nodes identified.

Gastrointestinal System

The esophagus is empty and unremarkable. The stomach contains approximately 100 cc of thick, bloody fluid and a small amount of unrecognizable food. There are no drug-like residues, pills, or capsules in the stomach. The stomach mucosa is congested with the usual rugal folds. The remainder of the gastrointestinal system is opened and is unremarkable, except for mucosal congestion.

The retroperitoneum is unremarkable.

Pancreas

The pancreas weighs 235 grams. The parenchyma is purple to brown and partially autolyzed.

Urogenital System

The kidneys are in the usual position and without malformation. The right kidney weighs 305 grams, and the left kidney weighs 295 grams. The surfaces are smooth and glistening. The capsules strip easily, revealing a red-brown surface. The corticomedullary junctions are well defined, though the parenchyma is edematous and mildly congested. The renal papillae have no hemorrhage or necrosis. The calyceal and collecting systems are not remarkable. The renal arteries and veins are unremarkable.

The ureters are not dilated or obstructed.

The bladder is empty. The bladder exhibits the usual tannish-pink mucosa with focal hemorrhage consistent with the presence of a Foley catheter. The ureteral orifices are patent.

The prostate is not enlarged and does not constrict the urethra. The tissue of the prostate is lobulated, tan, and moderately firm.

Adrenals

Both adrenals are of the usual size and shape. The cut surface shows a thin yellow cortex and brown-gray medulla, which are partially autolyzed.

Musculoskeletal System

There are no gross bony deformities. The muscles are well developed and of the usual color and consistency. The sternum, ribs, and spine exhibit the usual bone density and marrow.

Neck

The soft tissues of the neck, the thyroid and cricoid cartilages, larynx, and the hyoid bone show no hemorrhages or evidence of traumatic injury. The thyroid gland weighs 30 grams. The parenchyma is reddish brown and homogeneous. There are no paratracheal hemorrhages or masses. The epiglottis and vocal cords are unremarkable. The neck has been examined at the conclusion of the autopsy, after the blood has drained and the tissues are dry.

Central Nervous System

The scalp is reflected from mastoid process to mastoid process, revealing an approximately 10 × 4.5 cm area of mid to deep subcutaneous and subgaleal hemorrhage in the lateral left temporal parietal scalp with some hemorrhage in the fascia over the left temporalis muscle. There are small patchy areas of hemorrhage in the left temporalis muscle. The calvarium is intact, and when removed, there is no evidence of epidural or subdural hemorrhage. The dura mater is white and smooth and does not exhibit any stains or discolorations. The leptomeninges are not remarkable.

The brain weighs 1370 grams and a detailed neuropathological report will follow.

The dura covering the vault and the base of the cranium is removed.

The basilar skull is intact.

The atlanto-occipital articulation is intact. The odontoid process shows no fractures or dislocations. The cervical spine appears to be intact.

Note: Evidence obtained includes sample of pubic hair and fingernail clippings from both hands.

All evidence was collected by the autopsy technician and placed in an appropriately labeled envelope, with the name of the deceased. Evidence will be submitted to the crime lab.

Microscopic Examination

Heart
Sections show focal interstitial and perivascular myocardial fibrosis.

Lungs
Sections show extensive pulmonary vascular congestion with some postmortem extravasation of blood or antemortem intra-alveolar hemorrhage. Due to artifacts of processing, intra-alveolar edema cannot be appreciated. There is patchy to extensive infiltration of alveoli by macrophage, but no acute disease process is noted.

Trachea
Section shows patchy submucosal hemorrhage with no associated acute inflammation. The submucosal blood vessels are congested.

Liver
Sections show some of the hepatocytes contain small to large clear intracytoplasmic vacuoles consistent with mild fatty change. The portal tracts contain a mild to moderate amount of chronic inflammatory cells, some are expanded by fibrosis (with or without bile ductular hyperplasia present), and there is focal fibrous bridging between portal tracts. This may represent a nonspecific chronic portal triaditis or more likely a chronic persistent hepatitis (possibly hepatitis C viral infection in light of the fatty change), though a clinical correlation is required. The sinusoids are congested, predominantly

in the centrilobular areas. Focally these areas are quite hemorrhagic, and this may mask an acute centrilobular necrosis.

Pancreas

Section shows areas of fibrosis with isolated acini or islets consistent with chronic pancreatitis.

Kidneys

Sections show many tubules have degenerating/fragmenting cells lacking nuclei without any associated acute inflammation. This may represent early acute tubular necrosis (possibly due to terminal hypoxia) or a postmortem change.

Left Side of Scalp (Slides 1 and 2)

There are five pieces of subcutaneous soft tissues (i.e., admixture of fibrous tissue, fat, or muscle) with evidence of acute hemorrhage. In one section only is there, focally, admixed very mild to mild acute inflammation (i.e., neutrophils) within the hemorrhage.

Left Temporalis Muscle

The sections of muscle show small areas of acute hemorrhage within the muscle without any associated acute inflammation.

The microscopic examination is otherwise consistent with the gross findings and final pathological diagnoses.

Anatomic Diagnoses

    I. Positional and Mechanical Asphyxia

   II. Ischemic Encephalopathy

 III. Multisystem Organ Failure, Clinical, with:
   - A. Metabolic acidosis
   - B. Rhabdomyolysis
   - C. Disseminated intravascular coagulopathy
   - D. Respiratory failure
     - a. Mechanical ventilation dependent
     - b. Pulmonary congestion and edema, marked (right lung: 990 grams and left lung: 900 grams)
   - E. Acute renal failure
   - F. Hypokalemia
   - G. Hypotension
   - H. Hyperglycemia
   - I. Anasarca, with

   a. Bilateral pleural effusion (approximately 700 cc each side)
   b. Ascites (approximately 300 cc)
   c. Pericardial effusion (30 to 40 cc)
IV. Cardiomegaly (Heart Weight: 640 Grams) with Left Ventricular Hypertrophy
   A. Myocardial fibrosis, focal
 V. Evidence of Trauma
   A. Abrasions of nose, upper back, right wrist, right knee, and on/around left elbow
   B. Focal skin avulsion below right knee
   C. Subcutaneous and subgaleal hemorrhage of lateral left temporal parietal scalp
   D. Fascial hemorrhage over left temporalis muscle
   E. Left temporalis muscle hemorrhage, patchy
VI. Acute Alcohol Intoxication, Clinical (Serum Alcohol Level: 132 mg%)
VII. Fatty Change of Liver, Mild
VIII. Chronic Pancreatitis
  IX. Generalized Congestion of Viscera

Opinion

The decedent, a 43-year-old African American male, died as a result of positional and mechanical asphyxia, which occurred as he was being handcuffed while face down on the ground. Cardiomegaly with left ventricular hypertrophy is a contributory factor. The hypoxia/anoxia that occurred as he was being handcuffed most likely led to a cardiac arrhythmia, especially in light of his underlying heart disease. The combination of hypoxia/anoxia and cardiac arrhythmia then resulted in the development of ischemic encephalopathy and multiorgan system failure.

Manner of Death

The manner of death was homicide.

Autopsy Report 2—Positional Asphyxia Following TASER

External Examination

The body is that of a well-developed, well-nourished, muscular, African American male, whose appearance is consistent with the reported age of 37 years, measuring 5 feet 10 inches with an estimated antemortem body weight of approximately 185 pounds.

The body is clad in the following articles of apparel:

1. Black two-piece suit
2. White, long-sleeved, dress shirt
3. White, short-sleeved, T-shirt
4. Plastic mortuary undergarments
5. Black socks

The body shows the following evidence of previous autopsy and post-mortem embalming procedures:

1. Plastic cups overlie the eyes.
2. A bitemporal incision over the top of the scalp is closed tightly by thick, white string.
3. Two strings bind the upper and lower gums, one slightly to the right and one to the left of the midline.
4. A "Y"-shaped thoracoabdominal incision is closed tightly by thick, white string.
5. A previously made incision in the right lower thoracic area extending into the right upper abdominal quadrant is closed tightly by thick, white string. (We are informed by the funeral director at this time that this was not a postmortem trochar incision site.) A few abrasions are noted surrounding this sutured area.
6. An incision from the midoccipital portion of the cranium straight down in vertical fashion over the vertebral column, ending slightly below the belt line is closed tightly by thick, white string. Some cotton mesh is present overlying the sacrococcygeal region.
7. A previously made incision on the left posterolateral infrascapular region is closed by thick, white string. (We are informed by the funeral director at this time that this was not a postmortem trochar incision site.)

The body shows the following evidence of external injuries:

1. After the cosmetic material has been removed from the face, a diffuse area of dull, dark reddish brown discoloration is seen in the upper right lateral cheek area. It measures approximately 2 × 1.5 cm.
2. There is an area of dark, dull reddish-brown contusion on the left side of the chin. This is located 1/2 cm to the left of the midline. It measures 3 cm horizontally × 2.8 cm vertically.
3. There are three small areas of superficial, dull reddish-brown discoloration seen on the right side of the chin. These are present in staggered fashion, each measuring approximately 0.4 × 0.2 cm.

4. There is a large area of dull, dark reddish-black discoloration, representing a contusion on the anterior crest of the right shoulder. This measures approximately 7.5 cm horizontally × 5.5 cm vertically.

5. There is an area of somewhat darker reddish-brown discoloration, in which abrasions of the anterior lateral crest of the left shoulder are also noted. It has a slightly diagonal orientation. It measures 6.5 cm horizontally × 2.8 cm vertically in greatest dimensions.

6. There is an irregularly shaped area of abrasion-contusion on the dorsolateral aspect of the left wrist, extending slightly onto the proximal most portion of the left hand. This measures 2.5 cm vertically × 2.3 cm horizontally in greatest dimensions.

7. There is an area of abrasion-contusion on the dorsomedial aspect of the distal portion of the left forearm. This measures approximately 1 cm × 1 cm in greatest dimension. The inferior margin is located approximately 3.5-cm proximal to the left wrist area.

8. There is an area of circular abrasion-contusion on the lateral aspect of the knuckle of the left second finger, measuring 0.4 cm in diameter.

9. There is an irregularly shaped area of abrasion-contusion on the dorsal surface of the right hand. This measures approximately 4.1 × 2.1 cm in greatest dimensions. The superior most margin of this injury is located approximately 2.5 cm below the wrist level.

10. There is an irregularly shaped area of abrasion-contusion just proximal to the space between the knuckles of the right second and third fingers on the dorsal surface of the hand. This measures approximately 0.6 × 0.3 cm in greatest dimensions.

11. There is an area of abrasion-contusion at the base of the knuckle of the right thumb. This has a dark brownish-black appearance. It measures approximately 3 × 1 mm.

12. There is an area of abrasion-contusion on the superior surface of the right knee with a dark brownish-black appearance. This measures approximately 0.4 × 0.2 cm.

13. There is an area of superficial abrasion on the superior aspect of the left knee. This has a somewhat staggered configuration, measuring approximately 1.5 cm horizontally × 1 cm vertically in greatest dimensions.

14. There is a staggered area of superficial abrasions-contusions on the dorsolateral aspect of the proximal portion of the left forearm. These have a parallel streaking configuration. The largest component measures approximately 4.5 cm vertically × 1.5 cm horizontally. The smaller, thinner streaking areas are seen inferiorly and medially, and also laterally in the left elbow region. These are similar in color and configuration, but are smaller in area than the larger region of injury described above.

15. There are small superficial abrasions seen on the right forearm, one on the dorsomedial aspect of the right elbow, and smaller ones on the dorsomedial aspect of the proximal portion of the right forearm. The largest of these measures approximately 0.3 × 0.1 cm.
16. There is a superficial area of dark, dull reddish-brown ecchymosis in the right posterior parietal portion of the scalp. This measures 0.4 × 0.2 cm.
17. In the left upper forearm, extending downward from the antero-medial aspect of the left antecubital fossa, there is a diffuse area of darker discoloration. This measures approximately 10 cm vertically × 6.5 cm horizontally. Multiple incisions into this area disclose diffuse, extensive, dark red hemorrhage throughout the underlying subcutaneous tissues and superficial layers of the muscle bundles.

Remainder of External Examination

The scalp is completely devoid of hair. When the covering plastic cups are removed from the eyes, no well-defined petechial hemorrhages are seen. No foreign bodies, hemorrhages, or exudates are present within the auditory, nasal, or oral cavities. There does not appear to be any conjunctival suffusion or scleral icterus. The nasal septum is intact and in the midline. A hair growth of a couple to a few days is present in the nasolabial fold, and a small tuft of hair is seen beneath the midline of the lower lip. The lips, gums, and teeth are intact and show no evidence of injuries. The teeth appear to be in fairly good condition.

The neck shows no increased mobility on manipulation.

There is an old, well-healed scar in the right supraclavicular region with slight keloid formation. There is another old, well-healed scar in the right lateral infraclavicular area with keloid formation.

The abdomen is flat and firm. There is an old, well-healed scar running in diagonal fashion in the right lower abdominal quadrant, parallel to and just above the inguinal canal.

The pubic escutcheon and external genitalia have a normal adult male appearance. The penis has been circumcised. The testes appear to be palpable within the scrotal sac at this time.

The upper and lower extremities show no deformities such as to suggest recent fractures or dislocations. The hands, fingers, and fingernails show the injuries described above. A flat, dark brownish, pigmented nevus is seen on the volar aspect of the distal portion of the right forearm with a few tufts of hair. The fingernails appear to have been previously trimmed. Artefactual subungual hemorrhage is noted at the distal tips of the fingers (most probably associated with postmortem trimming and embalming procedures). The toenails are intact. The soles of the feet are clean. An old, well-healed scar is seen on the anterior aspect of the midportion of the left calf.

Rigor mortis is fixed. Livor mortis cannot be discerned due to the pigmentation of the skin and postmortem embalming.

The previously described incised wound in the left posterolateral lower back region is opened. It appears that a portion of skin has been excised in a somewhat circular fashion. This defect measures approximately 2 cm in diameter. The surrounding skin and underlying subcutaneous tissue are unremarkable.

The previously described incision in the right lower anterolateral chest wall is opened. This appears to be a previously excised, fairly circular area of skin and underlying tissue, measuring approximately 1.5 cm in diameter. The surrounding skin shows dark brown dyspigmentation.

Internal Examination

Head

The bitemporal incision is opened by cutting through the string. No subscalpular, subgaleal, or subperiosteal hemorrhages are noted. The calvarium is intact with no evidence of fractures or dislocations.

When the previously cleaved calvarium is lifted, there is no evidence of hemorrhage on the undersurface of the calvarium. The cranial vault is filled with cotton and solid white, powdery, paraformaldehyde material.

The posterior medial portions of both middle cranial fossae, extending across the middle of the basilar skull, reveal a dark reddish-brown to reddish-black discoloration. This is also seen in the anterior portions of both posterior cranial fossae, extending into the rim of the foramen magnum. All of these areas have a dark reddish-brown to reddish-black appearance, which appears to be artefactual.

There is subperiosteal hemorrhage of the sphenoid ridges. Incised portions of the sphenoid ridges in the base of the skull reveal evidence of hemorrhage into the underlying bone.

Vertebral Column and Spinal Cord

Hemorrhagic staining is noted along the lower thoracic and lumbar vertebrae.

The electric saw is used to remove portions of the vertebral column in the region of the staining. Examination of these pieces of removed bone reveals evidence of hemorrhage in the overlying periosteum. Hemorrhage does not extend into the vertebral bodies.

The midline incision of the back is opened, revealing an intact vertebral column with no evidence of fractures of the spinous processes or transverse processes. Subcutaneous hemorrhage is noted in the lower thoracic vertebral area, measuring approximately 5.5 × 2.7 cm.

The spinal cord is removed by means of an electric saw. No fractures of the vertebral bodies or any of their articular processes are noted.

The spinal cord is intact with no evidence of transection, laceration, or contusion. When the cord is removed, there appears to be subdural hemorrhage in the distal portion, extending into the cauda equina. This hemorrhage is dark, dull reddish-brown in color. No liquid blood is seen. This hemorrhage extends for a distance of approximately 7 cm.

No incisions are made into the cord at this time. It will be preserved in formalin for several days and then subsequently examined.

Oral Cavity

The oral cavity is examined. No injuries are noted. The tongue and surrounding tissues have been previously removed, extending into the hypopharyngeal region.

Neck

The neck area is examined. There are no fractures or dislocations of the cervical vertebral bodies or their processes. The entire laryngeal area was previously removed.

Internal Organs

The internal organs have been transmitted in a separate, large, plastic bag.

The heart cannot be identified.

Portions of the aorta reveal no evidence of atherosclerosis.

The lungs show focal areas of intraparenchymal hemorrhage and severe congestion. No infarcts, abscesses, tumors, or emphysematous changes are noted.

The tracheobronchial tree reveals no evidence of obstructions. The cartilaginous rings are intact. The hyoid bone and thyroid and cricoid cartilages are intact with no evidence of fractures or dislocations. No hemorrhages are seen in the surrounding soft tissues.

The spleen shows marked congestion. No lacerations, contusions, or infarcts are seen.

The liver shows no lacerations, tumors, or contusions. There is no gross evidence of fatty change or cirrhosis.

The pancreas shows no evidence of hemorrhage, inflammation, or necrosis. No tumors are seen.

The kidneys show no lacerations, contusions, infarcts, or tumors. The external surfaces are smooth and intact. Cut sections show normal corticomedullary architectures with no dilatation of the pelves or calyces. The renal papillae have no hemorrhage or necrosis. The ureteropelvic junctions are unobstructed and have no exudate or calculi.

The urinary bladder is unremarkable.

The prostate gland is unremarkable.

The intestines are unremarkable. There is no evidence of laceration, contusion, tumors, obstructions, or ulcers.

The distal esophagus and stomach are present. There is hemorrhage in the subserosal tissues of the distal esophagus at the esophageal-gastric junction. No ulcers are seen. The mucosal surface of the esophagus and stomach is intact with no evidence of ulcerations or tumors.

Various portions of the brain are present. These reveal no evidence of old or recent infarcts. No laceration, contusions, infarcts, tumors, or hemorrhages are seen. There is no evidence of any kind of vascular malformation.

Penis

The penile shaft is dissected horizontally and vertically, revealing extensive, bright red, fresh hemorrhage, extending from the dorsal to the ventral surface throughout more than half the length of the penis from the base outward.

The testes are intact with no evidence of hemorrhage or contusion. No hemorrhage is seen within the scrotal sac.

Final Pathological Diagnoses

Positional asphyxiation:
  Acute pulmonary congestion
  Acute pulmonary edema
Blunt force trauma and compression injuries of face and trunk:
  Contusions and abrasions of scalp, face, shoulders, arms, legs, and
    back, multiple subperiosteal hemorrhage, intercostal muscles
    and paravertebral soft tissues
Hemorrhage, sphenoid bones
Hemorrhage of penile shaft, extensive
Subdural hemorrhage, distal portion of spinal cord
Acute cocaine toxicity
Status, post-TASER impacts (×3)—clinical history

Sample Report of Opinions of a Coroner's Office

A fire hall was rented for a private birthday party that was attended by approximately 100 guests. The fire department required the individuals sponsoring the party to pay for security supplied by the police department. In this case, two police officers were mandated due to the number of attendees at the party. The two police officers, accompanied by the dog of one of the officers, were assigned by the police department. The primary purpose for the officers' presence was to maintain security and avoid any damage to the fire hall. Alcohol was served at this private party.

The fire hall was primarily composed of two large rooms on the second floor. One of the rooms was utilized as a bar room, and the second was a large room in which tables and chairs were erected for eating along with a buffet-style food service. During the time that food was being served, one of the men in the buffet line acted inappropriately by putting his hands into the spaghetti bowl rather than waiting to be served by one of the women standing behind the food table. At this point, the police became aware of this socially inappropriate conduct by the decedent's brother. The brother was not loud, boisterous, or threatening, although he was clearly lacking in socially acceptable conduct at that moment. Upon observing the situation, the officers approached the decedent's brother. According to the officers, he was noncommunicative and did not immediately respond to their inquiries. At this point, the decedent became aware of a situation involving his brother, and he proceeded to the location of the officers and his brother. The decedent attempted to determine the nature of the problem with his brother and to remove him from the area. The officers challenged him and refused to permit him to take charge of his brother. Words were exchanged between the decedent and the police which included swearing from both the decedent and the police. At that point, some of the attendees began to gather to view the situation. At no time did any of the guests come into physical contact with the police or verbally or physically threaten or assault them. The officers told the decedent that he was also under arrest, presumably for disorderly conduct. The decedent refused to accept the arrest and verbally challenged the police and began to back toward the door leading to the bar room. He did this at the urging of one of the women who was in charge of the party.

At or about this time, one of the officers called for backup, indicating the need for additional police assistance. In response, approximately 13 additional police officers representing both the borough and city police department appeared. An officer from the borough was the first to arrive. He approached the decedent, but instead of attempting to deescalate the situation, he followed the direction of one of the original officers on the scene to arrest the decedent. The decedent and the borough officer then exchanged words and shoving. During the period that followed, other officers arrived. The decedent was forced to the floor on his stomach, and he and a number of the police officers struggled until he was ultimately handcuffed behind his back. Although the length of time of the struggle differed from the various witnesses, it appears that the decedent was on the floor on his stomach anywhere from a minimum of five minutes to fifteen minutes.

The decedent weighed 300 pounds and was 6 feet tall. He had a large protruding abdomen. According to the non-police-officer witnesses, a number of the police officers who responded to the call for assistance subdued the decedent by exerting pressure on his upper torso by lying across his back while he laid on the floor on his stomach. The descriptions resembled a piling on with multiple police officers lying across the decedent's back, similar to a football piling on. According to the police officers' testimony, no officer placed any pressure on the decedent's back or upper torso during the event, except one officer who testified he placed one hand on one of the decedent's shoulders for a short period of time. The police officers who were physically engaged with the decedent indicated their attention was directed toward his legs or the lower portion of his body, and they were unaware of any activity that was occurring at the upper portion of his body. All of the other police officers offered no testimony on this issue, because they testified they were looking in the other direction and participated only in "crowd control." This totally and diametrically opposed testimony concerning the subduing of the decedent is most troubling. According to the nonpolice witnesses, the police were piling on the decedent as he lay on his stomach and was heard to exclaim he could not breathe. The police deny any piling-on activity occurred or that any of the police placed any pressure on the decedent's back.

It is clear from the testimony of the pathologist that the decedent died of mechanical asphyxiation that resulted from the inability of his body to supply oxygen to the brain. This happened as a result of pressure being applied to the decedent's back while he was lying on his stomach on the floor. The pressure caused the abdominal contents to compress his lungs, preventing them from expanding, thereby denying him the ability to breathe. This consequently denied the oxygenation of the blood and prevented oxygen from reaching the brain. As the doctor testified, there had to be force on the decedent's back while he lay on his stomach for a sufficient time to cause the life-threatening condition. The pathologist observed that even obese people roll over on their stomach while sleeping and do not die.

By the time the decedent arrived at the hospital that evening, he was "brain dead." He expired as a result of the conduct that occurred at the fire hall.

Neither the decedent's brother nor the decedent's conduct amounted to disorderly conduct at the time that the officers pronounced they were under arrest. It is the finding by the coroner's office that the decedent had not engaged in disorderly conduct as defined by Section 5503 of the Pennsylvania Crimes Code at the time he was placed under arrest. Therefore, the arrest of the decedent was without probable cause and was an illegal arrest.

Notwithstanding the illegality of the arrest, the law is well settled that a person may not resist an arrest even though it is unlawful and illegal. The individual must challenge the arrest in the court and not at the place where the arrest is effected. See Section 505 of the Pennsylvania Crimes Code.

However, under Section 505, a police officer, while effecting an arrest, whether legal or illegal, may not employ deadly force unless two elements are present:

1. The force is necessary to prevent the arrest from being defeated by resistance or escape.
2. The person has committed a forcible felony, is attempting to escape, and possesses a deadly weapon or otherwise indicates he or she will

endanger human life or inflict serious bodily injury unless arrested without delay.

The arrest of the decedent was not for a forcible felony, and he did not possess a deadly weapon or indicate he would endanger human life or inflict serious bodily injury unless arrested without delay.

It is therefore the opinion and finding of the coroner's office that the cause of death of the decedent was mechanical asphyxiation and the manner of death was homicide.

It is appropriate to comment further, that based upon the testimony, it appears impossible to suggest which of the police officers engaged in the prohibitive use of excessive force. None of the officers admit to engaging in the type of conduct described by the attendees at the party, and none of the attendees could identify the individual police officers who engaged in the prohibited conduct.

Summary

Six situations that law enforcement officers should be alert for to avoid positional/restraint asphyxia include:

1. Any extreme physical energy expenditure by a suspect.
2. Any physical interaction (wrestling to gain control) with a suspect.
3. All overweight suspects are at risk.
4. Any suspect who seems immune to pepper spray.
5. Any suspect who seems immune to shocks from a TASER or other electronic control device.
6. Any suspect that requires more than hand restraint (handcuffs).[4]

Endnotes

1. Donald T. Reay, Suspect Restraint and Sudden Death, *FBI Law Enforcement Bulletin*, May 1996.
2. IACP, *Transportation of Prisoners*, Concepts and Issues Paper, October 1, 1996, pages 3–4; IACP, *Arrests*, Concepts and Issues paper, June 2006, page 3.
3. Charly Miller, *Two Vital Tips to Help Law Enforcement Officers Avoid Causing Restraint Asphyxia*, 2001, page 2 (http://www.charlydmiller.com).
4. Ibid.

In-Custody Deaths

8

In-custody suicides typically occur in a police lockup or county jail, where prisoners are initially held immediately after arrest, and can be prevented by understanding the motivation for such suicides and reducing the opportunity for their occurrence.

There are over 3,300 municipal and county jails nationwide. These jails typically house unsentenced offenders, offenders already sentenced but waiting transfer to a state prison, and those sentenced to less than a year.

Not only are police agencies legally responsible for persons in their custody, the death of a prisoner can result in charges of misfeasance, malfeasance, or nonfeasance irrespective of the care that has been taken to safeguard the individual inmate.[1]

Since the implementation of the Death in Custody Reporting Act of 2000, data have been collected by the U.S. Department of Justice, Bureau of Justice Statistics, and analyzed. Now information is readily available to all jail and prison managers on what prisoners are at the greatest risk of death while in their custody.

In August 2005, the Bureau of Justice Statistics issued a special report with an analysis of jail and prison deaths during the time period 2000 through 2002. The analysis pointed out that:

- The suicide rate of inmates in local jails is over three times the rate in state prisons.
- The suicide rate of inmates in the nation's 50 largest jail systems is half that of other jails.
- Violent offenders in both local jails and state prisons had suicide rates over twice as high as those of nonviolent offenders.
- Inmates under the age of 18 had the highest suicide rate in local jails.
- Other than the youngest inmates, the suicide rate increased with inmate age with inmates between the ages of 18 and 24 being the least likely to commit suicide and those inmates over the age of 55 the most likely to commit suicide.
- White jail inmates were six times more likely to commit suicide than African American inmates and three times more likely to commit suicide than Hispanic inmates.
- Male inmates were 56% more likely to commit suicide than female jail inmates.

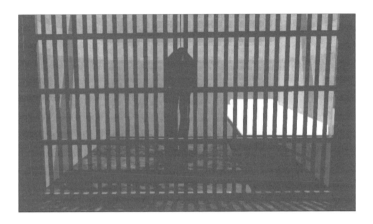

Figure 8.1 A jail suicide. (**See color insert following page 78.**)

- Violent offenders were three times more likely to commit suicide than nonviolent offenders.[2]

The data suggest that jail managers need to concentrate their suicide prevention efforts on the youngest and oldest of their inmates, on white males, and on violent offenders to effectively reduce jail suicides (Figure 8.1).

What Would You Have Done?

At approximately 2:00 P.M. on February 15, 2005, Jonathan Smith was arrested by a police officer employed by a medium-size northeastern police department. Smith, who was intoxicated at the time, was arrested at his residence for assaulting his wife, Jasmine, with a kitchen knife. Smith was transported to the police department where he was booked into the department jail by the shift supervisor, Lieutenant Buchannan. When screened by Buchannan before being placed into a cell, Smith told Buchannan that he was not thinking about suicide but had attempted suicide months before by taking pills, was taking antidepressants, and was under psychiatric care. After Smith was screened, photographed, and fingerprinted, he was placed into a cell.

On February 16, 2005, at approximately 9:00 P.M., jail officer Robert Jones heard a sound in the cell area and found Smith on his back on the floor in his cell. Jones immediately summoned jail Sergeant Lewis Albright to the area, and Albright ordered that a medical aide be called. Smith stated to Jones at that time that he had jumped off the bed and had hurt his head. After the medical aide checked Smith, a decision was made to transport Smith to a hospital to be checked for injuries.

On February 17, 2005, Smith was discharged from the hospital and transported back to the jail and was again placed in a cell; however, he was not placed on a special suicide watch as a prevention measure.

During the evening of February 17, 2005, Smith was discovered on the floor of his cell again. Lieutenant Buchannan went into the cell to check on Smith, and Smith told him that he would get off the floor in 10 minutes. Buchannan left him on the floor.

Buchannan checked on him again later, and because Smith refused to get up, he decided to lift him into a sitting position. Later, while jail officers were feeding the prisoners, Smith was observed by Buchannan sitting on his bench eating his food.

At approximately 9:50 p.m., a jail officer was giving the prisoners their prescribed medications and discovered Smith on the floor of his cell. When he rolled Smith on his side, he found phlegm around his mouth.

At approximately 10:00 p.m., Smith was transported by ambulance to a hospital in critical condition suffering from an acute subdural hematoma. He died the next morning.

Example—Pathological Examination

This 18-year-old male was found hanging in a holding cell in the police department. He fashioned a ligature from a belt of his trousers, wrapped it around a bar of the door of the holding cell and around his neck, and when discovered, it appeared as if he was standing slightly bent at the knees with his feet on the floor. The cell was opened and his body was cut down, falling to the floor. At this time, he had no pulse and he was unresponsive. An emergency medical services (EMS) squad was dispatched to the scene, and the deputy coroner arrived and pronounced him dead. The deceased was wearing very baggy pants and underwear. According to the police, the department received a call for a domestic dispute. The deceased was evidently fighting with his entire family, which prompted the call. He was placed under arrest by the officer and transported handcuffed with his hands behind his back to the police station for detention until paperwork was ready for his arraignment. He had been drinking beer during the day, and there was a history of cocaine use. Apparently, family members were shocked by the news of his actions, but they did state that he said he would never go back to jail again. He had been wearing a belt on his trousers when he was placed in the holding cell because the trousers were baggy and loose fitting and kept falling down. Investigation was also carried out by the district attorney's office.

Autopsy examination in this case was requested by the deputy coroner and the county coroner. The autopsy examination was to be complete. Family members were notified of the autopsy examination. The body was delivered to the hospital.

Any postmortem organ or tissue donation was declined because of the circumstances of the case.

Gross Description

General

The body was first viewed, partially clothed and covered by a white sheet, on the autopsy table in the necropsy room of the hospital for the preliminary autopsy examination. There is an orange and white identification bracelet filled with blue ink print with the name of the deceased present on the right wrist. No other means of identification are present, and color digital photos of the facial features were taken for identification purposes. The body was subsequently identified by the chief of the police department, who knew the deceased.

The body is resting straight on its back on the autopsy table with the arms and legs straight. The head is bent slightly at the neck and facing slightly toward the left side. At this time, there is advanced bilateral, symmetrical, and complete rigor mortis present in the muscles of the head and neck, jaw, and upper and lower extremities. There is marked purplish livor mortis of the posterior dependent portions of the body that blanches very slightly on pressure. There is striking contrasting pallor of the central back, buttocks, and forelegs. The body is cold to the touch (had been in refrigerated room).

External Evidence of Therapy

There are circular, white electrode pads with tabs adherent to the skin above the left breast and on the left side, and a clear circular electrode pad with a white tab adherent to the skin above the right breast.

Articles of Clothing

1. There is a pair of dirty, short, white sneakers with gray trim. They are about ankle length. Shoelaces are present. They are not tied at the top. At the top lace holes, the shoelaces are tied in a knot at the external aspect of the lace holes. There is reddish discoloration on the medial distal right sneaker above the sole. There is a label stating the make, size, and product numbers of the sneakers. There are several fine hairs inside the sneakers.
2. There is a pair of white terry cloth–like booties, about ankle length, with gray stitching about the toes and the heels. They are slightly dirty.
3. There is a pair of pale blue denim jeans with no belt. They are large sized and loose fitting for the body and are pushed or pulled down to the level of the mid thighs. They are heavy weight. They are dirty with areas of black smudging and discoloration. The waist button and the fly zipper are closed. The crotch area is slightly damp. There is a label

stating the brand, famous for superior product, crafted from the finest material, the size, 100% cotton, and product numbers. The left front, the left rear, the right rear, the right front, and watch pockets are empty. There are no defects noted in the slacks.

4. There is a pair of green and dark blue/black plaid boxer shorts. They are soft and flannel-like. There is a label stating the make, size, 50% cotton, 50% polyester. The crotch area is slightly damp.

5. Around the right upper arm, there is a loose-fitting, crimped, black elastic-like band with an athletic logo in white in two places and the brand name in one place in white print.

General

The body is that of a well-nourished, well-developed, muscular-appearing white male, looking his stated age of 18 years, measuring 73 1/2 to 74 in. in length and weighing an estimated 175 to 180 pounds. The body is tall and muscular appearing. The body had not been arterially embalmed. The hair is wavy, dark blonde, and measures about 3 to 3 1/2 in. in length. There are slight sideburns. There is a raspy growth of hair on the beard and mustache areas of the face and chin. There are mild diffuse hemorrhagic acne facial lesions throughout the face and neck area. The eyebrows are light brown. They are trimmed and moderately heavy. There are rare periorbital petechial hemorrhages and hemorrhages on the eyelids. The pupils are round and equal. They each measure 0.8 cm in diameter. The irises appear to be bluish or bluish-gray. The conjunctivae are congested. The bulbar conjunctivae are congested, but no definite petechial hemorrhages are evident. There are rare petechial hemorrhages evident in the lower tarsal conjunctivae of both eyes. The petechiae are not numerous. There is prominent congestion. There is no evidence of icterus. The ears, nose, and mouth are normal. There is no blood in the ears, nose, or mouth. The nasal septum is not perforated. There is moderate orodental hygiene and repair. There is grayish-brown discoloration of the left maxillary central incisor. The mandibular right and left canine teeth are pointed, and there is less pointing of the right and left maxillary canine teeth. The neck is symmetrical. There are no masses palpable. There is no abnormal mobility or crepitus elicited in the neck. The traumatic changes will subsequently be described. The chest is flat and firm. The muscles are prominent. The abdomen is flat and firm. The abdominal circumference varies from 32 to 34 in. The external genitalia are those of a normal adult male of stated 18 years of age. The penis is uncircumcised. The testicles are descended. The scrotal skin is somewhat dry and red. There is a moderate amount of grayish-white seminal-like fluid present at the urethral meatus and under the foreskin. There are no traumatic changes noted on the external genitalia. There is a linear area of abrasion or petechial hemorrhages in the right lower abdominal quadrant measuring 2.5 × 0.8 cm. There is no evidence of gravitational lividity of the lower abdomen, external genitalia, or distal arms or legs. There is hair on the forelegs, ankles, feet, and toes. There is no evidence of clubbing or edema. There are no "needle tracks" present on the skin of the arms. The fingernails are short. There is cyanosis of the nail beds of the fingers. In the skin of the left upper chest above the breast, there is a transverse, healed, keloidal type, pinkish-gray incision with evidence of suture marks resembling a scar measuring 8 × 1.1 cm. There is an oblique, similar appearing pinkish-gray keloidal-like scar in the posterior lateral aspect of the upper left arm and shoulder area, somewhat tapering, measuring 12 and up to 1.2 cm in greatest width, and a small scar about the lateral left elbow. Examination of the posterior aspect of the body reveals no unusual changes. The anal area appears unremarkable. Small lymph nodes are palpable in the inguinal area. Otherwise, there is no significant cervical, axillary, or inguinal lymphadenopathy.

External Evidence of Trauma

1. There are purplish-red ecchymoses with no swelling in the interphalangeal joint area of the right third and fourth fingers measuring 3 and 4 cm in greatest length. There is no evidence of crepitus. There is a brown ecchymotic area present on the dorsal aspect of the thenar web on the right hand, and in the area of the metacarpophalangeal joint area of the right third, fourth, and fifth fingers, there are tiny punctate and linear abrasions. There is no crepitus. There are circular reddish ecchymoses about the dorsal medial and lateral right wrist measuring about 1 to 2 cm in diameter.

2. There is a rectangular area of petechial hemorrhage and abrasion on the dorsal mid left forearm measuring 2.5 × 0.5 cm. There is an ill-defined area of yellowish-brown ecchymotic discoloration on the dorsal aspect of the distal left forearm. There is no crepitus in the upper or lower extremities or pelvic bones. There are small abrasions and ecchymoses about the lateral left elbow.

3. There is a dark red parchment-like abrasion with piling up of the epidermis on the lateral left hip measuring 3.2 cm in length and up to 1 cm in greatest width. There are multiple linear abrasions and abrasions and contusions about the skin of the left knee. Superior, there is a 3-cm dark red parchment-like abrasion. Inferior, there are two abrasions of similar appearance, together (measuring 6 × 1 cm). There are linear, smaller abrasions on the medial aspect. There are ill-defined contusions. There is an abrasion with scaliness and piling up of the epidermis on the proximal left foreleg. It is somewhat oblique (measuring 6 × 1 cm).

4. On the posterior aspect of the distal right foreleg, there are two band-like areas of blackish discoloration and smudging measuring 5 × 1.8 cm. There is a bluish ecchymotic discoloration on the proximal anterior right foreleg measuring 3 × 1.5 cm.

5. On the left hand at the interphalangeal joint area of the left middle and ring fingers, there are purple ecchymoses with no swelling and no crepitus measuring up to about 4 cm in greatest length and a 1 cm ecchymosis in a similar area on the little finger. At the metacarpophalangeal joint area of the third and fourth fingers, there are ecchymoses and small abrasions.

6. On the skin of the right lower dorsal back, there is an area of linear abrasions and scratches measuring about 8 × 2 cm. There is a superficial scratch or abrasion on the superior lateral left hip. It is small. On the lateral posterior left shoulder, there is a 1.2-cm dark red parchment-like abrasion. It is superficial with no underlying hemorrhage.

7. In the left temporoparietal occipital scalp, not evident without cutting the hair in the area, there is a roughly 7 × 5 cm area of small discrete circular ecchymoses and contusions with no swelling or lacerations or blood on the surface. The contusion resembles a patterned contusion. Subsequent examination reveals hemorrhage in the galea of the scalp, but no hematoma, extending to the periosteum.

8. Around the neck, there is an oblique ligature furrow with more tenting to the left side. On the anterior aspect of the neck, the ligature furrow is above the laryngeal prominence, and it is generally transverse but slightly oblique and extending on the right side. It is located 6 cm below the distal right earlobe when the measurement is taken to the top of the furrow, and this area measures 1.5 cm in width. There is a criss-crossing weave-like pattern in the base of the furrow corresponding to the pattern noted on the belt ligature to be subsequently described. The furrow is of no significant depth. On the left side, it ascends and is located 4 cm below the distal portion of the left earlobe when the measurement is taken to the top of the ligature furrow. On the right side, it extends posterior to the ear and ends in about the hairline. On the left side, it extends slightly into the hairline, but there is no circumferential ligature furrow about the neck. In the area of the left neck below and anterior to the earlobe around the area of the ramus of the mandible, there is a patterned area of pallor resembling a scalloped appearance or two semicircles of pallor with dark red outlines resembling an imprint of the belt buckle to be subsequently described measuring from about 1.5 to 2 cm in diameter. There are surrounding abrasions and ecchymoses. Parallel to the ligature furrow on the anterior and lateral aspects of the neck,

there are linear abrasions at the rim. They are reddish brown and parchment like. The ligature furrow has a yellow-brown to reddish appearance. A midline incision is made in the back from the occiput to the lower cervical spine. There is no evidence of fracture or soft tissue or skeletal muscle hemorrhage.

Submitted for examination by the deputy coroner is a cloth-like dark blue belt with two white metal buckles or loops. There is a large loop of belt that had been cut with one arm measuring 42 cm in length and the opposite arm measuring 50 cm in length. They measure 2.5 cm in width. Extending from the knot in the two buckles, there is a length of belt measuring 17 cm. There are adherent gray-white fragments on the belt resembling skin scales or lint. The belt is focally folded or creased longitudinally.

Internal Evidence of Trauma

The neck structures are dissected after examining the trunk and the central nervous system. There are congestive changes in the lower sternocleidomastoid muscles, but no definite hemorrhage. The strap muscles appear unremarkable with no hemorrhage. There is no definite hemorrhage about the larynx, trachea, or thyroid gland. The hyoid bone is not fractured. There is no hemorrhage. There is no hemorrhage about the arytenoid portions of the cornua of the laryngeal cartilage. The laryngeal mucosa appears unremarkable. There appears to be focal congestion or hemorrhage in the retropharyngeal area and about the proximal esophagus and between the esophagus and trachea. The common carotid arteries appear normal in size and coloration. They are not excised. A superficial adventitial, nontransmural incision reveals no evidence of intramural hemorrhage.

Primary Incision

A "Y"-type primary incision is made extending from the lateral clavicular areas to the symphysis pubis. The subcutaneous fatty and muscular tissues appear normal. The muscles are dark purplish-red, bulky, and have a good color and tonus. The adipose tissue measures up to 4 cm in thickness in the abdomen at a level below the umbilicus. There are no traumatic changes noted in the chest or abdominal wall. There is no evidence of fracture of the ribs or sternum.

Body Cavities

The pericardium is intact. The linings are smooth and glistening. The diaphragm reaches to the level of the fourth rib on the right and to the level of the fifth rib on the left. There was no escape of air noted when opening either pleural cavity. Both lungs are moderately distended and congested. They fill the entire pleural cavities, and their anterior medial borders are present in the anterior superior mediastinum over the pericardium, but they do not meet or overlap. Both sides of the diaphragm are intact. There is no excess fluid, adhesions, or traumatic changes noted in the peritoneal cavity. The peritoneal linings are smooth and glistening. There are no traumatic changes noted.

Heart

The heart weighs 405 grams. There are multiple epicardial petechial hemorrhages over the anterior lateral left ventricle and posterior lateral left ventricle and scattered on the anterior and posterior right ventricle. There are confluent petechial hemorrhages. Otherwise, the epicardium appears unremarkable. The apex of the heart is slightly pointed. On sectioning the heart, there is concentric left ventricular hypertrophy with narrowing of the chamber of the left ventricle. The left ventricle measures 2 cm in thickness. There is no gross evidence of acute or healed myocardial infarction. The right ventricle varies from 0.6 to 0.8 in thickness on the posterior aspect and outflow aspect. The endocardia of all cardiac chambers appear normal. There are no mural thrombi in any cardiac chamber, including the auricular appendages. There is postmortem fluidity of the blood in all chambers of the heart and great vessels. There is no evidence of postmortem clotting. All of the cardiac valves have a normal appearance. Their measurements in circumference are as follows: mitral valve, 10 cm; tricuspid valve, 14 cm; pulmonic valve, 8.5 cm; and aortic valve 7 cm. The coronary arteries arise normally behind the cusps of the aortic valve. There is no evidence of coronary ostial stenosis. The posterior descending artery is a continuation of the right coronary artery. There is a tiny accessory right coronary artery and ostium consistent with a conus artery. There are flat yellowish atherosclerotic streaks in the proximal left anterior descending and circumflex coronary arteries. Otherwise, there are no atherosclerotic changes present in any major epicardial coronary artery. The arteries are soft and pliable. The intimas are smooth and glistening. There is no evidence of severe stenosis, complete occlusion, or thrombosis. The arteries examined include the right coronary artery, left anterior descending coronary artery, major diagonal branch of the left anterior descending coronary artery, marginal branch of the circumflex left coronary artery, and the circumflex branch of the left coronary artery. The left main stem coronary artery measures 0.9 cm in length. There are no atherosclerotic changes present. There are no traumatic changes noted in the heart or thoracic vessels.

Lungs

The right lung weighs 930 grams, and the left lung weighs 800 grams. They are distended, congested, and purplish. On sectioning, there is moderate pulmonary congestion and edema involving all lobes of both lungs. There is no evidence of pulmonary emboli. The tracheobronchial system is intact with a small amount of tan mucoid fluid on the mucosal surface of the larynx, trachea, and major bronchi. There is no gross evidence of aspirated gastric contents or blood. Neck structures have been described above. There is no evidence of tongue bite. There are no foreign bodies noted in the oral cavity or oropharynx when palpating from the hypopharynx. There are no unusual changes noted in the anterior aspect of the cervical spine, including fracture or dislocation.

Mediastinum

There are no unusual gross changes noted. There is slight congestion. The thymus gland has a normal appearance for a male of stated 18 years of age and weight 50 grams.

Gastrointestinal System

The esophagus is normal. There is retroesophageal hemorrhage in the superior esophagus. The stomach is moderately dilated with a recently ingested, well-chewed meal with tannish fluid with greasy dark orange or yellow fluid on the surface mixed with innumerable fragments of food that are difficult to recognize. There appear to be small nodular fragments of stringy pale white meat and other fragments resembling red meat. There are numerous unrecognizable vegetable fragments and unrecognizable food fragments. There is no free fluid. There are no medication tablets evident. The mucosa appears normal. There is no evidence of tumor or ulceration. There is slight mucosal degeneration. The duodenum contains a small amount of cloudy, mucoid, yellowish-brown fluid. There is no evidence of tumor or ulceration. The lacteal pattern of the upper small intestine is extremely prominent. The network extends into the mesentery and into the proximal jejunal serosa. There are bile-like intestinal contents. Otherwise, the duodenum, jejunum, and ileum appear normal. They are slightly dilated, measuring up to 5 cm in diameter. The appendix is present. It appears normal except for a 1.5 fecalith in the proximal portion near its origin. There is no evidence of appendicitis. The cecum, ascending, transverse, descending, and rectosigmoid colon appear normal with slight dilatation of the ascending and transverse colon, measuring up to 6 cm in diameter. There is semifluid brownish fecal matter. Otherwise, the cecum, ascending, transverse, descending, and rectosigmoid colon appear normal with a small amount of soft, more formed fecal matter in the rectum. There are no tumors, polyps, or diverticula evident. The mesentery has a normal appearance.

Liver

The liver weighs 2,050 grams. The capsular surface is intact and brownish. There are no traumatic changes noted. On sectioning, there are moderate congestive changes. The gallbladder is normal and nondilated. There are no calculi. It contains 15 cc of a watery, dark yellow-brown bile. The common bile duct is normal and nondilated. There are no calculi in the common bile duct.

Pancreas

There are no unusual gross changes noted.

Lymph Nodes

There is no significant thoracic or abdominal lymphadenopathy. However, around the abdominal aorta near the bifurcation, there are prominent tan lymph nodes varying from 1 to 2 cm in greatest diameter. They are soft.

Spleen

The spleen weighs 200 grams. The capsular surface is intact and bluish gray. There are no traumatic changes noted. On sectioning, the pulp is dark purplish red. The follicles are not prominent. The pulp is slightly degenerated and semifluid consistent with postmortem change.

Kidneys

The right kidney weighs 170 grams, and the left kidney weighs 190 grams. The capsules strip with ease. The subcapsular surfaces are smooth and dark red with fetal lobulation. On sectioning, there are marked congestive changes. There is good corticomedullary demarcation. The right pelvis and ureter are normal and nondilated. There is slight dilatation of the left pelvis and ureter throughout that measure 1 cm in diameter.

Urinary Bladder and Internal Genitalia

The urinary bladder contains about 200 cc of a pale yellow urine. The bladder appears normal. The prostate gland is normal. The internal genitalia are those of a normal adult male of stated 18 years of age. The testicular tubules extend above the cut surface.

Endocrine System

The adrenal glands have a normal appearance and location. The thyroid gland is normal. It appears congested. One parathyroid gland grossly identified appears normal.

Cardiovascular System

The thoracoabdominal aorta and its major branches are normal with no atherosclerotic changes present. The inferior vena cava and iliac veins are normal. There are no thrombi in the inferior vena cava or iliac veins.

Musculoskeletal System

There is no evidence of fracture of the axial or appendicular skeleton, including the vertebral column and pelvic bones.

Central Nervous System

The scalp is reflected in the usual manner. In the left parieto-occipital galea scalp, there is an ecchymotic discolored area having a reddish appearance measuring 6.4 cm extending to the periosteum, but there is no hematoma or swelling. There is no evidence of fracture of the external vault of the skull. There are smaller circular ecchymoses in the right vertex and parietal area, each measuring about 2 cm in diameter. There is no thickness or hematoma. They are in the galea. There is ill-defined ecchymoses of the overlying scalp. In the temporal muscle on the left side near the ecchymosis described above, there is moderate subfascial hemorrhage in the temporal muscle, but no hematoma formation. The calvarium is removed in the usual manner. There is no evidence of fracture of the vault of the skull. It measures up to 0.8 cm in thickness. There is no evidence of epidural, subdural, or subarachnoid hemorrhage. There is marked meningovascular congestion. The brain and brain stem, together, weigh 1,460 grams. The cerebral hemispheres or supratentorial portions of the brain, together, weigh 1,260 grams. There is slight flattening of the gyri and narrowing of the sulci resembling cerebral edema or brain swelling with slight impaction of the cerebral hemispheres in the cranial cavity. Otherwise, on sectioning the cerebral hemispheres, cerebellum, brain stem, and proximal cervical spinal cord, there are no unusual gross changes noted. The vessels of the

Circle of Willis appear normal. The pituitary gland has a normal appearance. There is no evidence of basilar skull fracture. There is congestion in the ligaments of the atlanto-occipital joint, but no hemorrhage or lacerations. There are no unusual changes noted in the atlanto-occipital joint area or proximal cervical spinal canal.

Special Studies

1. Color digital photos of the autopsy findings were obtained.
2. Samples of all organs were placed in 4% formaldehyde to be submitted for histopathological examination.
3. Samples of vitreous fluid were removed from the right and left eyes. They were clear and colorless. The specimens will be submitted separately for sodium, potassium, chlorides, sugar, urea, and alcohol determinations.
4. Samples of blood were removed from the right and left inguinal areas to be submitted for blood typing, alcohol determination, carbon monoxide determination, drug screening for drugs of abuse, comprehensive toxicology blood drug screening, and blood thyroglobulin determination.
5. A sample of urine was aspirated from the urinary bladder by suprapubic puncture to be submitted for alcohol determination and drug screening. A Triage Plus PPX panel for drugs of abuse was positive for marijuana and cocaine.
6. A sample of gastric contents was retained to be submitted for alcohol determination.
7. A sample of bile was retained from the gallbladder to be submitted for drug screening and confirmation.
8. A nasal swab was obtained to be submitted for drug screening and confirmation.
9. X-rays of the skull and cervical spine were obtained in anterior-posterior and lateral views. The hyoid bone was x-rayed.

Cause of Death Based on Gross Findings at Conclusion of Autopsy and Pending Ongoing Official Investigative Findings, Toxicological and Histological Studies

The cause of death is hanging.

Manner of Death

The manner of death was suicide.

Medical Treatment of Prisoners

There can be no disagreement with the premise that every human being in a civilized society is entitled to basic medical care whenever possible and reasonably available in order to prevent that individual from dying. Deliberate, intentional withholding of medical care is anathema to the fundamental tenets of all major religions, and would constitute grounds for an official charge of homicide if a direct causal relationship were to be established between the failure to provide or acquire medical care for a critically ill person and the death. The degree of responsibility that any individual entity may have in such a situation will depend on several factors, one of which is the nature of the relationship between that individual and the person requiring medical care. To what extent does that involved individual have actual physical control over the sick person?

It is quite obvious and indisputable that when a police officer makes an arrest, the officer assumes control and custody of the person who has been arrested. It is also certain (with very few exceptions) that the arresting officer will have the ability to request assistance and additional input from other law enforcement officials if such backup is deemed necessary.

Even though a police officer is not expected or required to possess advanced medical training, knowledge, and skill, he is expected to comprehend and appreciate an obvious or reasonably apparent medical crisis and act accordingly. In the majority of such situations, the problem that the arrested person is suffering from will be readily perceived. Traumatic injuries, convulsions, bizarre behavior, verbalized complaints of severe physical pain, inability to respond in an appropriate manner to simple comments, loss of consciousness—these signs and symptoms of apparent, real, or possible physical or emotional distress must be looked for and responded to accordingly. Such evidence of an extant or developing medical problem should not be ignored. In some instances, the arrested person may be completely faking a medical problem. Other times, whatever medical problem does exist will prove to be of a relatively minor nature. However, such scenarios cannot be prematurely assumed. There will be no harm done if the officer overreacts to a troubling situation, and there subsequently proves to be nothing of a serious nature to contend with. On the other hand, a failure to appreciate that there is or could be a serious medical problem, and to deliberately or negligently delay obtaining medical care for a prisoner in distress, could lead to serious morbidity or death.

A significant percentage of people arrested for various kinds of offenses will be chronic alcoholics or drug addicts. Acute withdrawal from the abused drug, without any supportive medical care, can result in death. This well-known pathophysiological phenomenon must be understood and kept in mind in all arrest situations.

The police officer is not charged with the responsibility of undertaking heroic, life-saving, emergency medical procedures beyond whatever basic measures the officer may have been trained to perform during the course of his or her formal indoctrination program. However, there can be no acceptable rationale for failing to act quickly whenever an arrested individual appears to be in distress or is complaining of a significant medical problem. In such a situation, the sick person should be taken as quickly as possible to the nearest hospital emergency room or other medical facility if the hospital is too far away. If there is a dubious complaint, or a medical problem that is believed to be not too serious, it is wise to err on the conservative side. No harm will be done if the prisoner arrives at the jail an hour or more later because he or she had been taken first to a hospital for treatment or diagnostic evaluation.

Prior to locking up a newly arrested prisoner, background medical information should be obtained. Appropriate medical questionnaires are available for this purpose. This kind of form should be carefully and intelligently followed and completed before the prisoner is placed in a jail cell. If there are questions raised from the prisoner's responses, then the physician or nurse who has official responsibility and medical authority at that facility should be contacted as soon as possible and asked to see the prisoner or provide instructions as to what should be done.

Lives can be saved and lawsuits can be avoided if police officers are properly trained and thoroughly indoctrinated with the philosophy that all prisoners are human beings, and basic medical care is an inalienable right for all residents of our country.

Best Practices

The best practices to follow to prevent occurrences of suicide in local jails and prisons have been identified by the National Center on Institutions and Alternatives. Lindsay Hayes, in an article published in 2007, listed the key components to the prevention of suicides in jails:

1. A correctional staff that has been trained on obstacles to prevention; why correctional environments are conducive to suicide; potential predisposing factors to suicide; high-risk suicide periods; warning signs and symptoms; identifying suicidal inmates; components of an effective suicide prevention program; critical incident staff debriefing; and liability issues.
2. A good screening and assessment program when inmates enter the facility and an ongoing assessment of inmates at risk while in the facility.

3. Good communications when suicidal behavior is detected between transporting officer and correctional staff, between and among facility staff, between facility staff and the subject inmate, and between correctional staff and medical/mental health personnel.

4. Consideration of appropriate housing location, as physical isolation and restraint may be detrimental to the mental well-being of the inmate.

5. Increased visual supervision (either close observation or constant observation).

6. Swift intervention following a suicide effort.

7. Notification of all staff following a suicide effort.

8. A thorough follow-up and review after any suicide effort or inmate suicide to assess the circumstances surrounding the incident, the facility procedures relevant to the incident, the training that had been received by all relevant staff, pertinent medical and mental health services/reports involving the victim, possible precipitating factors leading to the suicide, and any recommendations for changes in policy, training, physical facility, medical or mental health services, and operational procedures.[3]

Endnotes

1. IACP National Law Enforcement Policy Center, *Lockups and Holding Facilities*, Concepts and Issues Paper, October 1, 1996.

2. Bureau of Justice Statistics Special Report, *Suicide and Homicide in State Prisons and Local Jails*, August 2005.

3. Lindsay M. Hayes, *Key Components of a Suicide Prevention Program*, National Center on Institutions and Alternatives, Baltimore, MD, 2007.

Emotionally Disturbed Persons

9

Since the deinstitutionalization of the mentally ill in the 1960s, responding to mentally ill people has become a large part of the police response to calls for their service. It is now a routine requirement that police officers assess the mental state and intentions of individuals they come into contact with while performing their duties. It is believed that a high percentage of homeless persons who live in public places on a part-time or full-time basis suffer from mental illness. Often, these people suffer from a combination of mental illness, alcoholism, and drug abuse.

The seminal study of police officer decision making regarding encounters with the mentally ill was Egon Bittner's study in 1967 on "Police Discretion in Emergency Apprehension of Mentally Ill Persons."[1] Bittner found that the police reluctantly made a psychiatric referral and initiated hospitalization only when the individual was causing serious trouble. Today, nothing has changed. According to a recent study, the police resolved 72% of all incidents involving mentally ill or emotionally disturbed persons by arrest and only initiated emergency hospitalization in 12% of the incidents.[2] Consequently, our jails have become the repository for people with mental disorders, and that will likely remain the case until law enforcement officers receive policy guidance and training on recognizing and properly handling the mentally ill.

According to some studies, between 10% and 25% of all shootings by the police involve suicide attempts.[3] Any shooting deliberately induced by a subject (intentionally forcing a confrontation with an officer that leaves the officer no choice but to use deadly force) is classified as an officer-assisted suicide or suicide-by-cop (SbC).[4]

Although some have suggested that suicidal individuals are not legitimate threats to officers, approximately 50% of the weapons used to threaten officers in SbC incidents are firearms, with the overwhelming majority being operative and loaded.[5]

Because of the increasing frequency of police officer encounters with the mentally ill and the difficulties they face when they encounter the emotionally disturbed and mentally ill, the International Association of Chiefs of Police (IACP) has now made a distinction between those who have committed a crime and those who become a focus of police intervention who have not committed a crime but are threatening suicide or are simply in a mental health crisis. Many of these persons will take a position in a vehicle or a structure and will refuse to exit and are legitimately classified as a "barricaded" person. Often, they will be armed with a knife or a firearm. In the past, the tactic employed when a person barricaded himself or herself was to call for a special tactical unit that would try first to negotiate with the barricaded person before efforts were undertaken to force the person from his or her shelter by physical force. Now, with the recognition that many of these people are mentally ill, and not criminals, the tactic has changed to the use of minimally intrusive techniques to resolve the situation.[6]

The IACP made it clear that dealing with mentally ill people was a real challenge for law enforcement officers when they stated in their 1997 Model Policy on the mentally ill the following:

> Dealing with individuals in enforcement and related contexts who are known or suspected to be mentally ill carries the potential for violence, requires an officer to make difficult judgments about the mental state and intent of the individual, and requires special police skills and abilities to effectively and legally deal with the person so as to avoid unnecessary violence and potential civil litigation. Given the unpredictable and sometimes violent nature of the mentally ill, officers should never compromise or jeopardize their safety or the safety of others when dealing with individuals displaying symptoms of mental illness. In the context of enforcement and related activities, officers shall be guided by this state's law regarding the detention of the mentally ill. Officers shall use this policy to assist them in defining whether a person's behavior is indicative of mental illness and dealing with the mentally ill in a constructive and humane manner.

Local sheriff's offices and police departments are statutorily tasked with the responsibility to take the mentally ill into protective custody for psychiatric evaluation, when a person demonstrates behavior that would lead a reasonable officer to believe that he or she is "gravely disabled" and unable to provide for basic human needs or presents a manifest danger to self or others. Additionally, the Americans with Disabilities Act (ADA) requires that the mentally ill be "reasonably accommodated" and that law enforcement contacts be conducted in such a manner as to not cause greater injury or indignity. Approximately 5% of the U.S. population has a serious mental illness, but many are functioning in a society that has adopted a public policy of discouraging institutionalization of those individuals who are responsive to therapy or treatment with psychotropic drugs. When such persons begin to decompensate and present a need for reevaluation, law enforcement officers will be called to actively intervene. Accordingly, an understanding of police interaction with the mentally ill in-crisis is essential to achieving custody without the avoidable violence that has too often resulted in injury or death.

The most important and fundamental issue to be kept in mind regarding the handling and treatment of emotionally disturbed individuals by police officers is the fact that most psychologically compromised people have not been identified and characterized as such at the time that the officer initially becomes involved. There is no simple, universal, and readily recognizable methodology that can be employed to enable law enforcement personnel to determine that the person who is creating a disturbance, or who may have committed an act of violence, is acting under the influence of alcohol, drugs, mental disorder, or a physiological disease process. Even a fully trained, experienced forensic psychiatrist or clinical psychologist would not be able to make such a rapid diagnosis. Therein is the critical problem in many such cases.

Inasmuch as serious harm to the individual in question, the police officer summoned or directed to the scene, family members or friends, or uninvolved third parties may occur if such an emotionally disturbed person has a potentially lethal weapon, it behooves the officer in charge to proceed with utmost care and caution. Many of these confrontations that terminate in human tragedy could be defused and handled without any violence if all law enforcement personnel were to be informed and trained to the fullest extent possible in how to *initially* recognize the *possibility* that the subject's erratic behavior and apparent refusal to respond in an appropriate fashion to the officer's instructions may be attributable to some kind of emotional disturbance. Unless the perceived or evolving situation is one that mandates immediate action, the officer should attempt to evaluate the problem and calmly determine if the actor appears to be in full control of his or her mental faculties and physical actions.

Alcohol, a great variety of psychotopic drugs, schizophrenia, an acute psychotic episode, and several clinicopathological processes (e.g., uncontrolled diabetes, epilepsy, and other convulsive seizure disorders, severe electrolyte imbalance, etc.) are all capable of causing a person to behave in a bizarre, uncontrollable, and potentially violent manner. What may seem to the involved officer to be outright defiance may in fact be a cerebral inability to understand and appreciate the circumstances of a confrontational dilemma and conceivable disaster.

All the usual precautionary measures and procedural conditions that are relevant and applicable to the disposition and treatment of medically ill individuals—while being arrested or during incarceration—must be kept in mind and rigidly adhered to in the handling of emotionally disturbed individuals. Such clinical scenarios as acute, unmonitored, untreated withdrawal from alcohol or drugs by an addict can lead to seemingly strange or even violent behavior and culminate in death. An unrecognized diabetic patient can react in an extremely bizarre fashion while in a hypoglycemic state. People suffering from grand mal seizures can die in several ways if their underlying problem is not recognized and treated.

All prison deaths, as well as any other death related to police involvement—scene altercation, pursuit, arrest, interrogation—should be reported to the local coroner or medical examiner. A complete, meticulous, well-documented postmortem examination, including toxicological analysis, should be performed by trained forensic scientists. Sometimes, in cases involving emotionally disturbed individuals, the cause of a sudden, unexpected death may not be definitive and unequivocal. Correlation of autopsy findings with the decedent's medical/hospital records, information gleaned from the scene investigation and interviews conducted by homicide detectives, and any other relevant observations must then be undertaken by the forensic pathologist in order to ascertain to the fullest extent possible what the pathophysiological mechanisms most likely were that resulted in the prisoner's death.

The manner of death in such cases could be natural, accidental, or homicide. Occasionally, the medical examiner or coroner may make an official ruling of "Undetermined" regarding either the cause or manner of death, or rarely, both. Obviously, it is imperative that all such fatalities be thoroughly, carefully, and objectively evaluated. The personal family and societal impact, and any possible ramifications of either or both a civil and criminal nature could be dependent upon these rulings.

What Would You Have Done?

This incident takes place in a formerly rural but rapidly developing county that is within commuting distance of a high-tech industrial employer and bordering a metropolitan area, where a mental health facility is located. The 34-person sheriff's office was experiencing an increase in calls for service, and while they occasionally had to take a mentally ill person into protective custody, they usually knew him or her. Recently, such calls had not only increased, but the persons they encountered were frequently strangers.

Sandra Jefferson was the 28–year-old only child of successful parents, who were both employed as researchers in the computer industry, and they had recently moved into the area for new job opportunities. Sandra had been diagnosed as bipolar in her early teens but was under a psychologist's care, had responded well to medication, and continued to live at home. She occasionally displayed mild depression or agitation, but her behavior was well within limits, and she had never threatened violence to herself or others. Her social life was limited, but she had recently been involved in a love affair with a younger neighborhood gardener, who had suddenly left the area without even telling her when or where he was going, and she soon became despondent. Unbeknownst to her parents, Sandra began flushing away her medications, instead of taking the prescribed dosage, and she gradually began to display greater mood swings than before. Because her parents were both employed and often worked long hours on new projects, they were slow to recognize the gradual change in their daughter's behavior.

One Sunday morning, both parents were awakened by Sandra's anguished wailing from their kitchen and rushed to see what had occurred. They soon found their daughter lying on the floor loudly sobbing and repeating over and over that she had no more reason to live. Her mother kneeled to comfort her and asked, "Sandra, what on earth is wrong?" The young woman answered that she was tired of living a hopeless existence and was going to kill herself. When both her mother and father asked the reason she would even think about doing such a thing, she excitedly replied, "I've always been a burden, nobody cares about me, I want to die!" Because neither parent had ever seen their daughter so distraught before, and it was obvious that she needed professional intervention, they called 911 for help.

The call taker at the sheriff's office functioned in the multicapacity role of dispatcher, clerk, and jail matron. She had been employed for less than a year, and her lack of seniority had relegated her to the 4:00 A.M. to 12:00 A.M. shift on Wednesday through Sunday mornings, when few events of any real significance ever occurred. Upon being told by the mother that her daughter was contemplating suicide, she only asked the caller's address and dispatched the single on-duty patrol unit to a "suicide-in-progress." The dispatched deputy was short on experience, too. He had never handled anything like this before, and he immediately asked that an off-duty sergeant be called out to assist. The dispatcher notified the assigned call-out sergeant and told him why the deputy was requesting his assistance. The dispatched deputy arrived 20 minutes before his sergeant and talked to both parents outside, who told him, "Our daughter is going to kill herself." After more conversation, the deputy learned that their daughter was bipolar, had a treating psychologist, was on medication but seemed to have been experiencing wider mood swings recently, and had never threatened suicide or violence to anyone before. The sergeant soon arrived, and the deputy relayed the information he had received from the parents. Telling the parents to wait outside, the two officers entered the home and found Sandra to still be in the kitchen, but she was standing and was holding a steak knife in each hand. When she saw the officers, she seemed surprised and asked them, "Why are you here and where are my parents?" Answering neither question but vividly remembering their recent in-service training on *Surviving Edged Weapons*, the sergeant immediately drew his weapon, pointed the gun at her, and repeatedly commanded in a loud voice, "Drop the knives and get down on the floor face down, now!" Sandra looked puzzled and did not respond, except to back into a corner of the kitchen still holding the two knives. With the officers blocking the only exit, the sergeant continued to hold her at gunpoint and sent the deputy outside to retrieve a 12-gauge shotgun loaded with beanbag rounds from the trunk of his patrol vehicle. When the deputy went outside, the worried parents asked what was going on, but he replied that he did not have time to talk, grabbed the shotgun, and rushed back into the house, while the parents shouted, "Don't hurt her, let us talk to her and call her doctor."

Sandra was still contained in the kitchen corner, with a knife in each hand, and, in spite of the sergeant continuing to give her loud commands, she was totally uncommunicative. The sergeant concluded that there was little to be gained from wasting more time, and he ordered the deputy to shoot her with the first beanbag round to cause her compliance. The deputy did as told, and from a distance of approximately 21 feet, he fired a round into Sandra's left thigh. She gave a startled cry and turned slightly to her left but remained standing and gripped the knives even more tightly. The next round struck her in the same area, with the same results, and, when the third round struck her left arm, she dropped the knife from that hand. The deputy was chambering the fourth round when Sandra suddenly screamed, raised the remaining knife over her head, and ran forward toward the officers in the kitchen doorway. The sergeant who was providing lethal cover to his deputy, rapidly fired five shots from his .40-caliber Glock, and Sandra went down with bullets in her chest and abdomen. Medics were called to the scene and found her to be deceased. The entire duration of the contact between the officers and the decedent was less than five minutes before she was killed.

The state police conducted an officer-involved shooting investigation for review by the district attorney, and she ruled the shooting to have been a justifiable homicide in self-defense.

Best Practices

Even though it has been a nationally recognized and accepted standard in the law enforcement profession since 1997, that law enforcement agencies provide their employees with guidance on how to deal with the mentally ill, it is clear that these officers were ill trained on how to deal with Sandra.[7]

Since 1996, state and local government agencies employing 50 or more persons have been required, by Title II of the Americans with Disabilities Act, to review their services, policies, and practices for compliance with the ADA (see ADA, 42 U.S.C. Section 12132) to insure individuals suffering from disabilities, such as Sandra, received proper services from government entities. However, the sheriff's department responding to Sandra's mental health crisis was too small to fall under the requirements of the ADA. Even though the Police Executive Research Forum (PERF) had published a Trainers Guide and Model Policy to assist local law enforcement agencies in developing their own policy statement on the proper response of law enforcement to people with mental illness, it is apparent that the officers responding had no such training.[8]

The protocol recognized and accepted in the law enforcement profession for dealing with the mentally ill when Sandra experienced a mental health crisis was to:

- Make every effort to take the person into protective custody for treatment.
- Attempt to calm the situation.
- Assume a nonthreatening manner.
- Move slowly so as not to excite the individual.
- Provide assurance that the police were there to help.
- Communicate with the individual to learn what was bothering him or her.
- Do not threaten the individual in any manner.
- Gather any helpful information about the individual from family members.[9]

Unfortunately, vividly remembering his recent in-service training on *Surviving Edged Weapons*, the sergeant immediately drew his weapon, pointed the gun at Sandra, and repeatedly commanded in a loud voice, "Drop the knives and get down on the floor face down, now." The threat resulted in Sandra closing on the officers with a knife in hand. At that point, deadly force was the predictable outcome.

Crises Intervention Teams

In 1987, a Memphis, Tennessee, police officer shot and killed a mentally ill man who had cut himself with a knife and was threatening others with the knife. The public outcry that followed the shooting in Memphis led to the formation of a community task force that developed a Crises Intervention Team. Even though it was recognized by both law enforcement professionals and mental health professionals that there would always be situations that would require the police to use lethal force, the creation of a Crises Intervention Team was based on the theory that the partnering of police and mental health professionals would lead to less violence and more effectiveness when interacting in the field with the mentally ill. Today, many law enforcement agencies have made operational Crises Intervention Teams in their agencies and, like Memphis, have continued to respond to a high level of calls involving the mentally ill, but have experienced fewer calls for hostage situations and barricaded subjects as officers have learned to deescalate situations by applying the training they received to be a member of the Crises Intervention Team.

Crime Scene and Forensic Evidence

Taking a mentally ill person into protective custody is often an extremely difficult task. There is a compelling need to act in a compassionate manner and to make every effort to not cause death or injury when attempting to take a person into protective custody for medical evaluation and treatment. At the same time, police officers justifiably are concerned that an uncooperative and unreasonable person in a mental health crisis can threaten their own safety. When officers use deadly force under circumstances similar to this scenario, the news coverage of such events is seldom positive (Figure 9.1).

In the scenario, officers were dispatched to handle a mentally ill person—Sandra Jefferson. The sergeant arrived to back up the deputy and ended up ordering Sandra to drop the knife. Sandra did not understand and failed to follow the order that led to the shooting death of Sandra. When this happens, the following essential tasks should be carried out:

- Secure the crime scene or institute crime scene protection measures. Because this is an indoor crime scene, the security of the scene is relatively simple.
- Remove the parents of Sandra from the scene and have an officer with special training or a victim advocate consult with the parents.
- Remove officer(s) involved in the shooting from the scene to avoid potential confrontation between the officers and Sandra's parents.
- Initiate normal crime scene investigation procedures.
- Notify appropriate agencies and people, including crime scene specialist, forensic laboratory, medical examiner, and district attorney's office.
- Media information for the department should release the department official statement related to the incident.

Figure 9.1 News articles on deaths related to police activity. (**See color insert following page 78.**)

General Scene Procedure

A general indoor crime scene search procedure should be followed. All the potential evidence, bullets, casings, and weapons should be secured in the original location and thoroughly documented before removal from the scene.

One of the common mistakes made in many shooting scenes is to dig out bullets and pick up evidence before the scene is thoroughly documented. This type of failure will make the subsequent reconstruction impossible.

As described in Chapter 3, the following procedures should be undertaken any time there is an officer-involved shooting (OIS).

Crime Scene Investigation

The crime scene investigation should be conducted by the agency's most experienced and best trained investigators. The location of bloodstains, bullets, casings, and weapons should be noted and documented. The position and location of the decedent also should be noted and documented.

The entire scene should be thoroughly documented using the standard crime scene documentation techniques of photography and sketches. Figure 9.2

Figure 9.2 A view of an indoor shooting scene. (**See color insert.**)

shows an overall view of an indoor shooting crime scene. The following areas are especially important to the successful reconstruction of the event:

- Photographs and measurement of bullet holes.
- Photographs and documentation of all patterns and wounds.
- Determination and documentation of bullet impact sites and their direction.
- Determination and documentation of gunshot residue (GSR) pattern evidence on the victim's body and clothing.
- Location and documentation of all spent bullets and casings.
- Location and documentation of all involved weapons.
- Photographs and documentation of all bloodstain patterns.

The Crime Scene Search

A combination of zone and link method could be used for this type of scene search. Because this scene is limited to a confined area, the number of crime scene investigators allowed to enter the scene should be limited to avoid accidental contamination or alteration of the scene.

Collection, Preservation, and Packaging of Physical Evidence

The following types of forensic evidence must be collected in all police-involved shootings:

1. Evidence from Police Officer
   a. Weapons and ammunition
   b. Hand swabs for GSR
   c. Clothing and shoes
   d. Statements related to incident
2. Evidence from Decedent
   a. Clothing and shoes
   b. Blood patterns and GSR patterns are extremely important and should be carefully preserved
   c. GSR hand swabs
   d. Gunshot wounds and other injury patterns
   e. Medical and autopsy reports
   f. Toxicology report
3. Evidence from Scene
   a. Bloodstains and their patterns
   b. Body position and location
   c. Guns, knives, and other weapons
   d. Bullets, casings, and fragments

e. Bullet trajectory
f. Ricochet and deflection patterns
g. GSR and GSR patterns
h. Markings, damages, and other patterns
i. Fingerprints, shoe prints, and other imprint evidence
j. Trace and transfer evidence such as hairs and fibers
k. Alcohol or drug containers

Preliminary Reconstruction

One of the most important aspects of this investigation is to reconstruct the shooting event. Reconstruction is based on the forensic and pattern evidence found at the scene. High-velocity blood spatters and the bullet holes could be used to determine the location and position of the decedent. The bullet trajectory and spent casing locations are useful information to determine officer position and location. GSR patterns found on the body and clothing could be used to estimate the distance between the weapons to the target. A more detailed reconstruction should be done after reviewing the autopsy results and forensic laboratory reports and evaluating witness observations and officer statements. This information should be useful to reconstruct the sequence of the event and to determine what, where, when, who, why, and how this shooting occurred. Reconstruction could also be used to confirm or refute the statements from witnesses, suspects, and police officers.

Releasing the Scene

Scenes located inside of a house do not require urgency; however, scenes that are outside and subject to environmental changes should be processed as quickly as possible to prevent the risk of evidence contamination resulting with weather changes.

Laboratory Analysis

Physical evidence collected from the crime scene should be submitted to a forensic science laboratory for further analysis and reconstruction. The following results may be obtained:

- Ammunition
  - Number of projectiles fired, number of projectiles found at the scene or in the body

- - Projectile identification: bullet, shot pellet, bean pellets, slug
  - Lands, caliber, gauge, shot size
  - Class characteristics: manufacturer, style, type of weapon fired from
  - Individual characteristics, striations, land and grooves marks
  - Damage and ricochet marks
- Trace evidence: blood, DNA, hair, tissues, fiber, wood, wall material

Cartridge Cases/Shells

- Number of spent casings/shells found at scene consistent with the number of shots fired. Each of the casings should be examined and identified.
- Manufacturer
- Caliber and load specifications
- Factory versus reloads
- Individual characteristics
- Firing-pin impression
- Breech face marks
- Chamber marks
- Extractor/ejector marks
- Primer: center versus rim fire
- Powder: black powder versus smokeless
- Wad, cup, and other fillers

Weapon Examination

- Functionality—proper firing mechanism
- Firing-pin, breech lock, ejector, extractor mechanism
- Loading mechanism
- Single, automatic, double barrel
- Magazine or cylinder
- Trigger pull
- Trace/fingerprints/DNA
- Ownership history

Gunshot Residues

Gunshot residues consist of unburned powder particles, primer residues, lubricants, and barrel residues. These materials may be found either on target surfaces or on an individual's hand that discharged a firearm. The source of GSR is different;

- GSR from muzzle
  - Powder particles on target surface such as skin, body, wound, clothing, object, surface
- GSR from cylinder gap
  - GSR particles on hands, clothing
- Gunshot residues generally can be identified by atomic absorption (AA) or scanning electron microscopy with energy-dispersive x-ray (SEM-EDAX) analysis. GSR patterns are usually identified by the following techniques:
  - Visual examination
  - Microscopic examination
  - Infrared (IR) photography
  - Lead particles ID
- GSR particles ID

Examination of Ricochet Bullets

Each projectile should be examined for the evidence of ricochet. This finding is valuable for trajectory reconstruction. Projectiles may impact different types of materials or surfaces, such as yielding surfaces of soil, sand, water, tissue, sheet metal, wood, drywall, or frangible surfaces such as cinder blocks, bricks, stepping stones, or nonyielding surfaces such as stone, concrete, steel, or heavy metal. When a projectile impacts on nonyielding surfaces, the following may occur:

- The projectile will lose some energy.
- The depart is usually different than the incident angle.
- The projectile usually has deformation.
- Trace transfer evidence will usually be present.
- The impact surface may play an important role.

Bloodstain Pattern and Tissues Examination

Blood, tissues, and hairs often may be recovered at a shooting scene. The identification of these materials can provide valuable information for reconstruction.

The following should be checked for the presence or absence of biological material:

- Tissue, blood deposits on officer's clothing.
- Tissue, blood deposits on decedent's clothing.
- Tissue, blood deposits on weapons.
- Blood flows and drips patterns on body.

- High velocity blood spatters on hands or wall.
- High and medium blood spatters on arms.
- Bloodstains and aerial deposits on furniture.
- Blood, tissue deposits on floor or wall.
- DNA typing of those blood and tissues.

Reconstruction

There are many types of reconstruction. In this particular case, the following type of reconstruction should be conducted:

- Wound pattern analysis
- Distance estimation
- Blood spatter pattern analysis
- GSR pattern determination
- Gunshot wound pattern analysis
- Trajectory determination
- Casing ejection pattern analysis
- Bullet damage and ricochet pattern
- Location and position of police officer
- Location and position of decedent
- The sequence of the shooting event

Endnotes

1. Linda Teplin, Police Discretion and Mentally Ill Persons, *National Institute of Justice Journal*, July 2000.
2. Ibid.
3. Studies by Dr. Karl Harris, former Deputy Medical Examiner, Los Angeles County, California, 1983.
4. IACP Training Keys 535 and 536 entitled *Officer Assisted Suicide,* 2001, page 37.
5. Ibid.
6. IACP Model Policy entitled *Barricaded Subjects*, September 2007.
7. IACP Model Policy entitled *Dealing with the Mentally Ill*, April 1, 1997.
8. Police Executive Research Forum (PERF), Trainers' Guide and Model Policy entitled *The Police Response to People with Mental Illness*, 1997.
9. IACP Model Policy entitled *Dealing with the Mentally Ill*, April 1, 1997, page 1.

Index